THE 50th Law

50 CENT AND ROBERT GREENE

harperstudio

An Imprint of HarperCollins*Publishers*

HarperCollins books may be purchased for educational, business, or sales promotional use. For information please write: Special Markets Department, HarperCollins Publishers, 10 East 53rd Street, New York, NY 10022.

FIRST EDITION

Designed by Leah Carlson-Stanisic

Library of Congress Cataloging-in-Publication Data has been applied for.

ISBN 978-0-06-177460-7

09 10 11 12 13   OV/RRD   10 9 8 7 6 5 4 3 2 1

Foreword

I first met 50 Cent in the winter of 2006. He had been a fan of my book *The 48 Laws of Power*, and he was interested in collaborating on a book project. In the meeting, we talked about war, terrorism, the music business. What struck me most was that we had a remarkably similar way of looking at the world, one that transcended the great differences in our backgrounds. For instance, in discussing the power games he was experiencing at that time in the music business, we both looked past people's benign explanations for their behavior and tried to figure out what they were really up to. He developed this way of thinking on the dangerous streets of Southside Queens where it was a necessary life skill; I came to it by reading a lot of history and observing the crafty maneuvers of various people in Hollywood, where I worked for many years. The perspective, however, is the same.

We left the meeting that day with an open-ended idea about a future project. As I pondered the possible theme of this book over the following months, I became increasingly intrigued by the idea of bringing our two worlds together. What excites me about America is its social mobility, people continually rising from the bottom to the top and altering the culture in the process. On another level, however, we remain a nation that lives in social ghettos. Celebrities generally congregate around other celebrities; academics and intellectuals are cloistered in their worlds; people like to associate with those of their kind. If we leave these narrow worlds, it is usually as an observer or tourist of another way of life. What seemed an interesting possibility here was to ignore our surface differences as much as possible and collaborate on the level of ideas—illuminating some truths about human nature that go beyond class or ethnicity.

With an open mind and the idea of figuring out what this book could be, I hung out with Fifty throughout much of 2007. I was given almost complete access to his world. I followed him on numerous high-powered business meetings, sitting quietly in a corner and observing him in action. One day I witnessed a raucous fistfight in his office between two of his employees, with Fifty having to personally break it up. I observed a fake crisis that he manufactured for the press for publicity purposes. I followed him as he mingled with other stars, friends from the hood, European royalty, and political figures. I visited his childhood home in Southside Queens, hung out with his friends from his hustling days, and got a sense of what it could be like to grow up in that world. And the more I witnessed him in action on all these fronts, the more it struck me that Fifty was a walking, living example of the historical figures I had written about in my three books. He is a master player at power, a kind of hip-hop Napoleon Bonaparte.

While writing about the various power players in history, I developed the theory that the source of their success could almost always be traced to one single skill or unique quality that separated them from others. For Napoleon, it was his remarkable ability to absorb a massive amount of detail and organize it in his mind. This allowed him to almost always know more than his rival generals about what was going on. After observing Fifty and talking to him about his past, I decided that the source of his power is his utter fearlessness.

This quality does not manifest itself in yelling or obvious intimidation tactics. Any time Fifty acts that way in public it is pure theater. Behind the scenes, he is cool and calculating. His lack of fear is displayed in his attitude and his actions. He has seen and lived through too many dangerous encounters on the streets to be remotely fazed by anything in the corporate world. If a deal is not to his liking, he will walk away and not care. If he needs to play a little rough and dirty with an adversary, he goes at it without a second thought. He feels supreme confidence in himself. Living in a world where most people are generally timid and conservative, he always has the advantage of being willing to do more, to take risks, and to be unconventional. Coming from an environment in which he never expected to live past the age of twenty-five, he feels like he has nothing to lose, and this brings him tremendous power.

The more I thought of this unique strength of his, the more it seemed inspiring and instructive. I could see myself benefiting from his example and overcoming my own fears. I decided that fearlessness in all its varieties would be the subject of the book.

The process for writing *The 50th Law* was simple. In observing and talking to Fifty, I noticed certain patterns of behavior and themes that would eventually turn into the ten chapters of this book. Once I determined these themes, I discussed them with him, and together we shaped them further. We talked about overcoming the fear of death, the ability to embrace chaos and change, the mental alchemy you can effect by thinking of any adversity as an opportunity for power. We related these ideas to our own experiences and to the world at large. I then expanded on these discussions with my own research, combining the example of Fifty with stories of other people throughout history who have displayed the same fearless quality.

In the end, this is a book about a particular philosophy of life that can be summed up as follows—your fears are a kind of prison that confines you within a limited range of action. The less you fear, the more power you will have and the more fully you will live. It is our hope that *The 50th Law* will inspire you to discover this power for yourself.

The 50th Law

Introduction

SO OVER YOU IS THE GREATEST ENEMY A MAN CAN HAVE
AND THAT IS FEAR. I KNOW SOME OF YOU ARE AFRAID
TO LISTEN TO THE TRUTH—YOU HAVE BEEN RAISED ON
FEAR AND LIES. BUT I AM GOING TO PREACH TO YOU
THE TRUTH UNTIL YOU ARE FREE OF THAT FEAR. . . .

—Malcolm X

The Fearful Attitude

In the beginning, fear was a basic, simple emotion for the human animal. We confronted something overwhelming— the imminent threat of death in the form of wars, plagues, and natural disasters—and we felt fear. As for any animal, this emotion had a protective function—it allowed us to take notice of a danger and retreat in time. For us humans, it served an additional, positive purpose—we could remember the source of the threat and protect ourselves better the next time. Civilization depended on this ability to foresee and forestall dangers from the environment. Out of fear, we also developed religion and various belief systems to comfort us. Fear is the oldest and strongest emotion known to man, something deeply inscribed in our nervous system and subconscious.

Over time, however, something strange began to happen. The actual terrors that we faced began to lessen in intensity as we gained increasing control over our environment. But instead of our fears lessening as well, they began to multiply in number. We started to worry about our status in society—whether people liked us, or how we fit into the group. We became anxious for our livelihoods, the future of our families and children, our personal health, and the aging process. Instead of a simple, intense fear of something powerful and real, we developed a kind of generalized anxiety. It was as if the thousands of years of feeling fear in the face of nature could not go away—we had to find something at which to direct our anxiety, no matter how small or improbable.

In the evolution of fear, a decisive moment occurred in the nineteenth century when people in advertising and journalism discovered that if they framed their stories and appeals with fear, they could capture our attention. It is an emotion we find hard to resist or control, and so they constantly shifted our focus to new possible sources of anxiety: the latest health scare, the new crime wave, a social faux pas we might be committing, and endless hazards in the environment of which we were not aware. With the increasing sophistication of the media and the visceral quality of the imagery, they have been able to give us the feeling that we are fragile creatures in an environment full of danger—even though we live in a world infinitely safer and more predictable than anything our ancestors knew. With their help, our anxieties have only increased.

Fear is not designed for such a purpose. Its function is to stimulate powerful physical responses, allowing an animal to retreat in time. After the event, it is supposed to go away. An animal that cannot not let go of its fears once the threat is gone will find it hard to eat and sleep. We are the animal that cannot get rid of its fears and when so many of them lay inside of us, these fears tend to color how we view the world. We shift from feeling fear because of some threat, to having a fearful attitude towards life itself. We come to see almost every event in terms of risk. We exaggerate the dangers and our vulnerability. We instantly focus on the adversity that is always possible. We are generally unaware of this phenomenon because we accept it as normal. In times of prosperity, we have the luxury of fretting over things. But in times of trouble, this fearful attitude becomes particularly pernicious. Such moments are when we need to solve problems, deal with reality, and move forward, but fear is a call to retreat and retrench.

This is precisely what Franklin Delano Roosevelt confronted when he took office in 1933. The Great Depression that had begun with the stock market crash of 1929 was now at its worst. But what struck Roosevelt was not the actual economic factors but the mood of the public. It seemed to him that people were not only more fearful than necessary but that their fears were making it harder to surmount adversity. In his inaugural address to the country, he said that he would not ignore such obvious realities as the collapse of the economy and that he would not preach a naive optimism. But he implored his listeners to remember that the country had faced worse things in its past, periods such as the Civil War. What had brought us out of such moments was our pioneer spirit, our determination and resolve. This is what it means to be an American.

Fear creates its own self-fulfilling dynamic—as people give in to it, they lose energy and momentum. Their lack of confidence translates into inaction that lowers confidence levels even further, on and on. "So, first of all," he told the audience, "let me assert my firm belief that the only thing we have to fear is fear itself—nameless, unreasoning, unjustified terror, which paralyzes needed efforts to convert retreat into advance."

What Roosevelt sketched out in his speech is the knife's edge that separates failure from success in life. That edge is your attitude, which has the power to help shape your reality. If you view everything through the lens of fear, then you tend to stay in retreat mode. You can just as easily see a crisis or problem as a challenge, an opportunity to prove your mettle, the chance to strengthen and toughen yourself, or a call to collective action. By seeing it as a challenge, you will have converted this negative into a positive purely by a mental process that will result in positive action as well. And in fact, through his inspiring leadership, FDR was able to help the country shift its mind-set and confront the Depression with a more enterprising spirit.

Today we seem to face new problems and crises that test our national mettle. But just as FDR made the comparison to even worse times in the past, we can say that what we are facing is not as bad as the perils of the 1930s and the subsequent war years. In fact, the reality of twenty-first-century America is something more like the following: our physical environment is safer and more secure than any other moment in our history. We live in the most prosperous country in the world. In the past only white males could play the power game. Now millions upon millions of minorities and women have been given entrance to the arena, forever altering the dynamic—making us the most socially advanced country in that regard. Advances in technology have opened up all kinds of new opportunities; old business models are dissolving, leaving the field wide open for innovation. It is a time of sweeping change and revolution.

We face certain challenges as well. The world has become more competitive; the economy has undeniable vulnerabilities and is in need of reinvention. As in all situations, the determining factor will be our attitude, how we *choose* to look at this reality. If we give in to the fear, we will give disproportionate attention to the negative and manufacture the very adverse circumstances that we dread. If we go the opposite direction, cultivating a fearless approach to life, attacking everything with boldness and energy, then we will create a much different dynamic.

Understand: we are all too afraid—of offending people, of stirring up conflict, of standing out from the crowd, of taking bold action. For thousands of years our relationship to this emotion has evolved—from a primitive fear of nature, to generalized anxiety about the future, to the fearful attitude that now dominates us. As rational, productive adults we are called upon to finally overcome this downward trend and to evolve beyond our fears.

The Fearless Type

THE VERY FIRST THING I REMEMBER IN MY EARLY CHILD-HOOD IS A FLAME, A BLUE FLAME JUMPING OFF A GAS STOVE SOMEBODY LIT. . . . I WAS THREE YEARS OLD. . . . I FELT FEAR, REAL FEAR, FOR THE FIRST TIME IN MY LIFE. BUT I REMEMBER IT ALSO LIKE SOME KIND OF ADVENTURE, SOME KIND OF WEIRD JOY, TOO. I GUESS THAT EXPERIENCE TOOK ME SOMEPLACE IN MY HEAD I HADN'T BEEN BEFORE. TO SOME FRONTIER, THE EDGE, MAYBE, OF EVERYTHING POSSIBLE . . . THE FEAR I HAD WAS ALMOST LIKE AN INVITATION, A CHALLENGE TO GO FORWARD INTO SOMETHING I KNEW NOTHING ABOUT. THAT'S WHERE I THINK MY PERSONAL PHILOSOPHY OF LIFE . . . STARTED, WITH THAT MOMENT. . . . IN MY MIND I HAVE ALWAYS BELIEVED AND THOUGHT SINCE THEN THAT MY MOTION HAD TO BE FORWARD, AWAY FROM THE HEAT OF THAT FLAME.

—Miles Davis

There are two ways of dealing with fear—one passive, the other active. In the passive mode, we seek to avoid the situation that causes us anxiety. This could translate into postponing any decisions in which we might hurt people's feelings. It could mean opting for everything to be safe and comfortable in our daily lives, so no amount of messiness can enter. When we are in this mode it is because we feel that we are fragile and would be damaged by an encounter with the thing we dread.

The active variety is something most of us have experienced at some point in our lives: the risky or difficult situation that we fear is thrust upon us. It could be a natural disaster, a death of someone close to us, or a reversal in fortune in which we lose something. Often in these moments we find an inner strength that surprises us. What we feared is not so bad. We cannot avoid it and have to find a way to overcome our fear or suffer real consequences. Such moments are oddly therapeutic because finally we are confronting something real— not an imagined fear scenario fed to us by the media. We can let go of this fear. The problem is that such moments tend to not last very long or repeat themselves too often. They can quickly lose their value and we return to the passive, avoidance mode.

When we live in relatively comfortable circumstances, the environment does not press on us with obvious dangers, violence, or limitations to our movement. Our main goal then is to maintain the comfort and security we have, and so we become more sensitive to the slightest risk or threat to the status quo. We find it harder to tolerate feelings of fear because they are more vague and troubling—so we remain in the passive mode.

Throughout history, however, there are people who have lived in much tighter circumstances, dangers pressing in on them on a daily basis. These types must confront their fears in the active mode again and again and again. This could be growing up in extreme poverty; facing death on the battlefield or leading an army in war; living through tumultuous, revolutionary periods; being a leader in a time of crisis; suffering personal loss or tragedy; or having a brush with death. Countless people grow up in or with such circumstances and their spirit is crushed by adversity. But a few rise above. It is their only positive choice—they must confront these daily fears and overcome them, or submit to the downward pull. They are toughened and hardened to the point of steel.

Understand: no one is born this way. It is unnatural to not feel fear. It is a process that requires challenges and tests. What separates those who go under and those who rise above adversity is the strength of their will and their hunger for power.

At some point, this defensive position of overcoming fears converts to an offensive one—a fearless attitude. Such types learn the value not only of being unafraid but also of attacking life with a sense of boldness and urgency and an unconventional approach, creating new models instead of following old ones. They see the great power this brings them and it soon becomes their dominant mind-set.

We find these types in all cultures and all time periods— from Socrates and the Stoics to Cornelius Vanderbilt and Abraham Lincoln.

Napoleon Bonaparte represents a classic fearless type. He began his career in the military just as the French Revolution exploded. At this critical moment in his life, he had to experience one of the most chaotic and terrifying periods in history. He faced endless dangers on the battlefield as a new kind of warfare was emerging, and he navigated through innumerable political intrigues in which one wrong move could lead to the guillotine. He emerged from all of this with a fearless spirit, embracing the chaos of the times and the vast changes going on in the art of war. And in one of his innumerable campaigns, he expressed the words that could serve as the motto for all fearless types.

In the spring of 1800 he was preparing to lead an army into Italy. His field marshals warned him that the Alps were not passable at that time of year and told him to wait, even though waiting would spoil the chances for success. The general replied to them, "For Napoleon's army, there shall be no Alps." And mounted on a mule, Napoleon proceeded to personally lead his troops through treacherous terrain and past innumerable obstacles. It was the force of one man's will that brought them through the Alps, catching the enemy completely by surprise and defeating them. There are no Alps and no obstacles that can stand in the way of a person without fears.

Another example of the type would have to be the great abolitionist and writer Frederick Douglass, who was born into slavery in Maryland in 1817. As he later wrote, slavery was a system that depended on the creation of deep levels of fear. Douglass continually forced himself in the opposite direction. Despite the threat of severe punishment, he secretly taught himself to read and write. When he was whipped for his rebellious attitude, he fought back and saw that he was whipped less often. Without money or connections, he escaped to the North at the age of twenty. He became a leading abolitionist, touring the North and telling audiences about the evils of slavery. The abolitionists wanted him to stay on his lecture circuit and repeat the same stories over and over, but Douglass wanted to do much more and he once again rebelled. He founded his own antislavery newspaper, an unheard-of act for a former slave. The newspaper went on to have tremendous success.

At each stage of his life Douglass was tested by the powerful odds against him. Instead of giving in to the fear—of whippings, being alone on the streets of unfamiliar cities, facing the wrath of the abolitionists—he raised his level of boldness and pushed himself further onto the offensive. This confidence gave him the power to rise above the fierce resistances and animosities of those around him. That is the physics that all fearless types discover at some point—an appropriate ratcheting up of self-belief and energy when facing negative or even impossible circumstances.

Fearless people do not emerge exclusively from poverty or a harsh physical environment. Franklin Delano Roosevelt grew up in a wealthy, privileged family. At the age of thirty-nine he contracted polio, which paralyzed him from the waist down. This was a turning point in his life, as he faced a severe limitation to his movement and possibly an end to his political career. He refused, however, to give in to the fear and the downward pull on his spirit. He went the opposite direction, struggling to make the most of his physical condition and developing an indomitable spirit that would transform him into our most fearless president. For this type of person, any kind of encounter with adversity or limitation, at any age, can serve as the crucible for forging the attitude.

The New Fearless Type

THIS PAST, THE NEGRO'S PAST, OF ROPE, FIRE, TORTURE . . . DEATH AND HUMILIATION; FEAR BY DAY AND NIGHT, FEAR AS DEEP AS THE MARROW OF THE BONE . . . THIS PAST, THIS ENDLESS STRUGGLE TO ACHIEVE AND REVEAL AND CONFIRM A HUMAN IDENTITY . . . YET CONTAINS, FOR ALL ITS HORROR, SOMETHING VERY BEAUTIFUL. . . . PEOPLE WHO CANNOT SUFFER CAN NEVER GROW UP, CAN NEVER DISCOVER WHO THEY ARE . . . .

—James Baldwin

Through much of the nineteenth century, Americans faced all kinds of dangers and adversity—the hostile physical environment of the frontier, sharp political divisions, a lawlessness and chaos that came out of great changes in technology and social mobility. We responded to this constrictive environment by overcoming our fears and developing what came to be known as a pioneer spirit—our sense of adventure and our renowned ability to solve problems.

With our growing prosperity this began to change. In the twentieth century, however, one environment remained as harsh as ever—the black ghettos of inner-city America. And out of such a crucible a new fearless type came forward, exemplified by such figures as James Baldwin, Malcolm X, and Muhammad Ali. But the racism of the times constricted their ability to give full rein to this spirit.

In recent times, newer types have emerged from inner-city America with more freedom to advance to the highest points of power in America—in entertainment, politics, and business. They come from a Wild West–like environment in which they have learned to fend for themselves and give full rein to their ambition. Their education comes from the streets and their own rough experiences. In a way, they are throwbacks to the freewheeling types of the nineteenth century, who had little formal schooling but created a new way of doing business. Their spirit fits the disorder of the twenty-first century. They are fascinating to watch and in some ways have much to teach us.

The rapper known as 50 Cent (aka Curtis Jackson) would have to be considered one of the more dramatic contemporary examples of this phenomenon and this type. He grew up in a particularly violent and tense neighborhood—Southside Queens in the midst of the crack epidemic of the 1980s. And in each phase of his life he has had to face a series of dangers that both tested and toughened him, rituals of initiation into the fearless attitude he has slowly developed.

One of the greatest fears that any child has is that of being abandoned, left alone in a terrifying world. It is the source of our most primal nightmares. This was Fifty's reality. He never knew his father, and his mother was murdered when he was eight years old. He quickly developed the habit of not depending on other people to protect or shelter him. This meant that in every subsequent encounter in life in which he felt fear, he could turn only to himself. If he did not want to feel the emotion, he had to learn to overcome it—on his own.

He began hustling on the streets at any early age, and there was no way he could avoid feeling fear. On a daily basis he had to confront violence and aggression. And seeing fear in action so routinely, he understood what a destructive and debilitating emotion it could be. On the streets, showing fear would make people lose respect for you. You would end up being pushed around and more likely to suffer violence because of your desire to avoid it. You had no choice—if you were to have any kind of power as a hustler, you had to overcome this emotion. No one could read it in your eyes. This meant that he would have to place himself again and again in the situations that stimulated anxiety. The first time he faced someone with a gun, he was frightened. The second time, less so. The third time, it meant nothing.

Testing and proving his courage in this way gave him a feeling of tremendous power. He quickly learned the value of boldness, how he could push others on their heels by feeling supreme confidence in himself. But no matter how tough and hardened they become, hustlers usually face one daunting obstacle—the fear of leaving the streets that are so familiar and that have taught them all of their skills. They become addicted to the lifestyle, and even though they are likely to end up in prison or die an early death, they cannot leave the hustling racket.

Fifty, however, had greater ambitions than to become merely a successful hustler, and so he forced himself to face and overcome this one powerful fear. At the age of twenty and at the peak of his hustling success, he decided to cut his ties to the game and dive into the music racket without any connections or a safety net. Because he had no plan B, because it was either succeed at music or go under, he operated with a frantic, bold energy that got him noticed in the rap world.

He was still a very young man when he had faced down some of the worst fears that can afflict a human—abandonment, violence, radical change—and he had emerged stronger and more resilient. But at the age of twenty-four, on the eve of the release of his first record, he came face-to-face with what many of us would consider the ultimate fear—that of death itself. In May of 2000 an assassin poured nine bullets into him in broad daylight as he sat in a car outside his house, one bullet going through his jaw and coming within a millimeter of killing him.

In the aftermath of the shooting, Columbia Records dropped him from the label, canceling the release of his first album. He was quickly blackballed from the industry, as record executives were afraid to have any kind of involvement with him and the violence he was associated with. Many of his friends turned against him, perhaps sensing his weakness. He now had no money; he couldn't really return to hustling after turning his back on it, and his music career seemed to be over. This was one of those turning points that reveals the power of one's attitude in the face of adversity. It was as if he were confronting the impassable Alps.

At this moment, he did as Frederick Douglass did—he decided to ratchet up his anger, energy, and fearlessness. Coming so close to death, he understood how short life could be. He would not waste a second. He would spurn the usual path to success—working within the record industry, nabbing that golden deal, and putting out the music they thought would sell. He would go his own way—launching a mix-tape campaign in which he would sell his music or give it away for free on the streets. In this way he could hone the hard and raw sounds that he felt were more natural to him. He could speak the language of the hood without having to soften it at all.

Suddenly he felt a great sense of freedom—he could create his own business model, be as unconventional as he desired. He felt like he had nothing to lose, as if the last bits of fear that still remained within him had bled out in the car that day in 2000. The mix-tape campaign made him famous on the streets and caught the attention of Eminem, who quickly signed Fifty to his and Dr. Dre's label, setting the stage for Fifty's meteoric rise to the top of the music world in 2003, and the subsequent creation of the business empire he has forged since.

We are living through strange, revolutionary times. The old order is crumbling before our eyes on so many levels. And yet in such an unruly moment, our leaders in business and politics cling to the past and the old ways of doing things. They are afraid of change and any kind of disorder.

The new fearless types, as represented by Fifty, move in the opposite direction. They find that the chaos of the times suits their temperament. They have grown up being unafraid of experimentation, hustling, and trying new ways of operating. They embrace the advances in technology that make others secretly fearful. They let go of the past and create their own business model. They do not give in to the conservative spirit that haunts corporate America in this radical period. And at the core of their success is a premise, a Law of Power that has been known and used by all the fearless spirits in the past and is the foundation of any kind of success in the world.

The 50th Law

THE GREATEST FEAR PEOPLE HAVE IS THAT OF BEING THEMSELVES. THEY WANT TO BE 50 CENT OR SOMEONE ELSE. THEY DO WHAT EVERYONE ELSE DOES EVEN IF IT DOESN'T FIT WHERE AND WHO THEY ARE. BUT YOU GET NOWHERE THAT WAY; YOUR ENERGY IS WEAK AND NO ONE PAYS ATTENTION TO YOU. YOU'RE RUNNING AWAY FROM THE ONE THING THAT YOU OWN—WHAT MAKES YOU DIFFERENT. I LOST THAT FEAR. AND ONCE I FELT THE POWER THAT I HAD BY SHOWING THE WORLD I DIDN'T CARE ABOUT BEING LIKE OTHER PEOPLE, I COULD NEVER GO BACK.

−50 Cent

The 50th Law is based on the following premise: We humans have generally little control over circumstances. People intersect our lives, doing things directly and indirectly to us, and we spend our days reacting to what they bring. Good things come our way, followed by bad things. We struggle as best we can to gain some control, because being helpless in the face of events makes us unhappy. Sometimes we succeed, but the margin of control that we have over people and circumstance is depressingly narrow.

The 50th Law, however, states that there is one thing we can actually control—the mind-set with which we respond to these events around us. And if we are able to overcome our anxieties and forge a fearless attitude towards life, something strange and remarkable can occur—that margin of control over circumstance increases. At its utmost point, we can even create the circumstances themselves, which is the source of the tremendous power that fearless types have had throughout history. And the people who practice the 50th Law in their lives all share certain qualities—*supreme boldness, unconventionality, fluidity,* and *a sense of urgency*—that give them this unique ability to shape circumstance.

A bold act requires a high degree of confidence. People who are the targets of an audacious act, or who witness it, cannot help but believe that such confidence is real and justified. They respond instinctively by backing up, by getting out of the way, or by following the confident person. A bold act can put people on their heels and eliminate obstacles. In this way, it creates its own favorable circumstances.

We are social creatures, and so it is natural for us to want to conform to the people around us and the norms of the group. But underneath this is a deep fear—that of sticking out, of following our own path no matter what people think of us. The fearless types are able to conquer this fear. They fascinate us by how far they go with their unconventionality. We secretly admire and respect them for this; we wish we could act more like they do. Normally it is hard to hold our attention; we shift our interest from one spectacle to the next. But those who fearlessly express their difference compel our attention on a deeper level for a longer duration, which translates into power and control.

Many of us respond to the shifting circumstances of life by trying to micromanage everything in our immediate environment. When something unexpected happens, we become rigid and we respond by employing some tactic that worked in the past. If events change quickly, we are easily overwhelmed and lose control. Those who follow the 50th Law are not afraid of change or chaos; they embrace it by being as fluid as possible. They move with the flow of events and then gently channel them in the direction of their choice, exploiting the moment. Through their mind-set, they convert a negative (unexpected events) into a positive (an opportunity).

Having a brush with death, or being reminded in a dramatic way of the shortness of our lives, can have a positive, therapeutic effect. Our days are numbered and so it is best to make every moment count, to have a sense of urgency about life. It could end at any moment. The fearless types usually gain such awareness through some traumatic experience. They are energized to make the most of every action, and the momentum this gives them in life helps them determine what happens next.

It is all rather simple: when you transgress this fundamental law by bringing your usual fears into any encounter, you narrow your options and your capacity to shape events. Your fear can even bring you into a negative field where your powers are reversed. Being conservative, for instance, can force you into a corner in which you are more likely to lose what you have in the long run because you also lose the capacity to adapt to change. Trying so hard to please people can actually end up pushing them away—it is hard to respect someone who has such an ingratiating attitude. If you are afraid to learn from your mistakes, you will more than likely keep repeating them. When you transgress this law, no amount of education, connections, or technical knowledge can save you. Your fearful attitude encloses you in an invisible prison, and there you will remain.

Observing the 50th Law creates the opposite dynamic—it opens possibilities, brings freedom of action, and helps create a forward momentum in life.

The key to possessing this supreme power is to assume the active mode in dealing with your fears. This means entering the very arenas you normally shy away from: making the very hard decisions you have been avoiding, confronting the people who are playing power games with you, thinking of yourself and what you need instead of pleasing others, making yourself change the direction of your life even though such change is the very thing you dread.

You deliberately put yourself in difficult situations and you examine your reactions. In each case, you will notice that your fears were exaggerated and that confronting them has the bracing effect of bringing you closer to reality.

At some point you will discover the power of *reversal*—overcoming the negative of a particular fear leads to a positive quality—self-reliance, patience, supreme self-confidence, and on and on. (Each of the following chapters will highlight this reversal of perspective.) And once you start on this path, it is hard to turn back. You will continue all the way to a bold and fearless approach to everything.

Understand: you do not have to grow up in Southside Queens or be the target of an assassin to develop the attitude. All of us face challenges, rivals, and setbacks. We choose to ignore or avoid them out of fear. It is not the physical reality of your environment that matters but your mental state, how you come to deal with the adversity that is part of life on every level. Fifty *had* to confront his fears; you must *choose* to.

Finally, your attitude has the power of shaping reality in two opposite directions—one that constricts and corners you in with fear, the other that opens up possibilities and freedom of action. It is the same for the mind-set and spirit that you bring to reading the chapters that follow. If you read them with your ego out in front, feeling that you are being judged here, or are under attack—in other words, if you read them in a defensive mode—then you will needlessly close yourself off from the power this could bring you. We are all human; we are all implicated by our fears; no one is being judged. Similarly, if you read these words as narrow prescriptions for your life, trying to follow them to the letter, then you are constricting their value—their application to your reality.

Instead you must absorb these words with an open and fearless spirit, letting the ideas get under your skin and affect how you see the world. Do not be afraid to experiment with them. In this way, you will shape this book to your circumstances and gain a similar power over the world.

IN MY VIEW . . . IT IS BETTER TO BE IMPETUOUS THAN CAUTIOUS, BECAUSE FORTUNE IS A WOMAN, AND IF YOU WISH TO DOMINATE HER YOU MUST BEAT HER AND BATTER HER. IT IS CLEAR THAT SHE WILL LET HERSELF BE WON BY MEN WHO ARE IMPETUOUS RATHER THAN BY THOSE WHO STEP CAUTIOUSLY.

—Niccolò Machiavelli

CHAPTER I

See Things for What They Are— Intense Realism

REALITY CAN BE RATHER HARSH. YOUR DAYS ARE NUMBERED. IT TAKES CONSTANT EFFORT TO CARVE A PLACE FOR YOURSELF IN THIS RUTHLESSLY COMPETITIVE WORLD AND HOLD ON TO IT. PEOPLE CAN BE TREACHEROUS. THEY BRING ENDLESS BATTLES INTO YOUR LIFE. YOUR TASK IS TO RESIST THE TEMPTATION TO WISH IT WERE ALL DIFFERENT; INSTEAD YOU MUST FEARLESSLY ACCEPT THESE CIRCUMSTANCES, EVEN EMBRACE THEM. BY FOCUSING YOUR ATTENTION ON WHAT IS GOING ON AROUND YOU, YOU WILL GAIN A SHARP APPRECIATION FOR WHAT MAKES SOME PEOPLE ADVANCE AND OTHERS FALL BEHIND. BY SEEING THROUGH PEOPLE'S MANIPULATIONS, YOU CAN TURN THEM AROUND. THE FIRMER YOUR GRASP ON REALITY, THE MORE POWER YOU WILL HAVE TO ALTER IT FOR YOUR PURPOSES.

The Hustler's Eye

THIS IS LIFE, NEW AND STRANGE; STRANGE, BECAUSE
WE FEAR IT; NEW, BECAUSE WE HAVE KEPT OUR EYES
TURNED FROM IT. . . . MEN ARE MEN AND LIFE IS LIFE,
AND WE MUST DEAL WITH THEM AS THEY ARE; AND IF WE
WANT TO CHANGE THEM, WE MUST DEAL WITH THEM IN
THE FORM IN WHICH THEY EXIST.

—Richard Wright

As a boy, Curtis Jackson (aka 50 Cent) had one dominant drive—ambition. He wanted more than anything the very things that it seemed he could never have—money, freedom, and power.

Looking out on the streets of Southside Queens where he grew up, Curtis saw a grim, depressing reality staring him in the face. He could go to school and take it seriously, but the kids who did that didn't seem to get very far—a life of low-paying jobs. He could turn to crime and make his money fast, but the ones who went for that either died young or spent much of their youth in prison. He could escape it all by taking drugs—once you start down that path there is no turning back. The only people he could see who led the life that he dreamed of were the hustlers, the drug dealers. They had the cars, the clothes, the lifestyle, the degree of power that matched his ambitions. And so by the age of eleven he had made the choice to follow that path and become the greatest hustler of them all.

The further he got into it, however, the more he realized that the reality was much grimier and harsher than he had imagined. The drug fiends, the customers, were erratic and hard to figure out. The fellow hustlers were all fighting over the same limited number of corners and they'd stab you in the back in an instant. The big-time dealers who ran the neighborhood could be violent and heavy-handed. If you did too well, someone would try to take what you had. The police were everywhere. One wrong move could land you in prison. How could he possibly succeed amid this chaos and avoid all of the inevitable dangers? It seemed impossible.

One day he was discussing the troublesome aspects of the game with an older hustler named Truth, who told him something he would never forget. Don't complain about the difficult circumstances, he said. In fact, the hard life of these streets is a blessing if you know what you're doing. Because it is such a dangerous world, a hustler has to focus intensely on what's going on around him. He has to get a feel for the streets—

who's trouble, where there might be some new opportunity. He has to see through all the bullshit people throw at him— their games, their lousy ideas. He has to look at himself, see his own limitations and stupidity. All of this sharpens the eye to a razor's edge, making him a keen observer of everything. That's his power.

The greatest danger we face, he told Curtis, is not the police or some nasty rival. It's the mind going soft. I've seen it happen to many a hustler, he said. If things go well, he starts thinking it will go on forever and he takes his eyes off the streets. If things go bad, he starts wishing it were all different and he comes up with some foolish scheme to get quick, easy money. Either way, he falls fast. Lose your grip on reality on these streets and you might as well kill yourself.

In the months to come, Curtis thought more and more about what Truth had told him, and it began to sink in. He decided to transform the hustler's words into a kind of code that he would live by: he would trust no one; he would conceal his intentions, even from friends and partners; and no matter how high or low life brought him, he would remain the supreme realist, keeping his hustler's eye sharp and focused.

Over the next few years he became one of the savviest hustlers in his neighborhood, operating a small crew that brought him good money. The future looked promising, but a moment's inattention got him trapped in a police sting, and at the age of sixteen he was sentenced to nine months in a shock rehabilitation center in upstate New York. In this unfamiliar space and with time to reflect, suddenly the words of Truth came back to him. This was not the time to get depressed or to dream, but to fix that hustler's eye on himself and the world he lived in. See it as it is, no matter how ugly.

He had unbridled ambition; he wanted real power, something he could build on. But no street hustler lasts that long. It's a young man's game. By the time hustlers reach their twenties, they slow down and something bad happens or they go scurrying into a low-paying job. And what blinds them to this reality is the money and lifestyle in the moment; they think it will go on forever. They're too afraid to try something else. It doesn't matter how clever you are—there's a ceiling to how high you can rise.

He had to wake up and get out while he was still young and his ambitions could be realized. He would not be afraid. And so based on these reflections, he decided he would make a break into music. He would find a mentor, someone who could teach him the ropes. He would learn everything he could about music and the business. He would have no plan B—it was either make it there or die.

Operating with a kind of desperate energy, he made the transition into music, carving a place for himself by creating a sound that was hard driving and reflected the realities of the streets. After a relentless mix-tape campaign in New York he got the attention of Eminem, and a record deal followed. Now he seemed to have realized his childhood ambitions. He had money and power. People were nice to him. Everywhere he went they flattered him, wanting to be a part of his success. He could feel it happening—the good press, the sycophantic followers—it was all starting to go to his head and dull his vision. On the surface everything looked great, but what was the reality here? Now more than ever he needed that clear, penetrating eye to see past all the hype and glamour.

The more he looked at it, the more he realized that the reality of the music business was as harsh as the streets. The executives who ran the labels were ruthless. They distracted you with their charming words, but in fact they could care less about your future as an artist; they wanted to suck you dry of every dollar they could get out of you. Once you were no longer so hot, you would find yourself slowly pushed to the side; your decline would be all the more painful for having once tasted success. In truth, you were a pawn in their game. A corner hustler had more power and control over his future than a rapper did.

And what about the business itself? Record sales were falling because people were pirating music or buying it in different forms. Anyone with two eyes could see that. The old business model had to go. But these very same executives who seemed so sharp were afraid to confront this reality. They held on tightly to the past and would bring everyone down with them.

Not Fifty. He would avoid this fate by moving in a different direction. He would forge a diversified business empire, music merely being a tool to get there. His decisions would be based on his intense reading of the changing environment that he had detected in music but was infecting all levels of business. Let others depend on their MBAs, their money, and their connections. He instead would rely upon that hustler's eye that had brought him from the bottom of America to the top in just a few short years.

The Fearless Approach

REALITY IS MY DRUG. THE MORE I HAVE OF IT, THE MORE POWER I GET AND THE HIGHER I FEEL.

—50 Cent

You might imagine that the streets that molded Fifty and the code he created for himself have little to do with your circumstances, but that is merely a symptom of your dreaming, of how deeply you are infected with fantasies and how afraid you are to face reality. The world has become as grimy and dangerous as the streets of Southside Queens—a global, competitive environment in which everyone is a ruthless hustler, out for him- or herself.

Truth's words apply to you as much as to Fifty: the greatest danger you face is your mind growing soft and your eye getting dull. When things get tough and you grow tired of the grind, your mind tends to drift into fantasies; you wish things were a certain way, and slowly, subtly, you turn inward to your thoughts and desires. If things are going well, you become complacent, imagining that what you have now will continue forever. You stop paying attention. Before you know it, you end up overwhelmed by the changes going on and the younger people rising up around you, challenging your position.

Understand: you need this code even more than Fifty. His world was so harsh and dangerous it *forced* him to open his eyes to reality and never lose that connection. Your world seems cozier and less violent, less immediately dangerous. It makes you wander and your eyes mist over with dreams. The competitive dynamic (the streets, the business world) is in fact the same, but your apparently comfortable environment makes it harder for you to see it. Reality has its own power—you can turn your back on it, but it will find you in the end, and your inability to cope with it will be your ruin. Now is the time to stop drifting and wake up—to assess yourself, the people around you, and the direction in which you are headed in as cold and brutal a light as possible. Without fear.

Think of reality in the following terms: the people around you are generally mysterious. You are never quite sure about their intentions. They present an appearance that is often deceptive—their manipulative actions don't match their lofty words or promises. All of this can prove confusing. Seeing people as they are, instead of what you think they should be, would mean having a greater sense of their motives. It would mean being able to pierce the facade they present to the world and see their true character. Your actions in life would be so much more effective with this knowledge.

Your line of work is another layer of reality. Right now, things might seem calm on the surface, but there are changes rippling through that world; dangers are looming on the horizon. Soon your assumptions about how things are done will be outdated. These changes and problems are not immediately apparent. Being able to see through to them before they become too large would bring you great power.

The capacity to see the reality behind the appearance is not a function of education or cleverness. People can be full of book knowledge and crammed with information but have no real sense of what's going on around them. It is in fact a function of character and fearlessness. Simply put, realists are not afraid to look at the harsh circumstances of life. They sharpen their eye by paying keen attention to details, to people's intentions, to the dark realities hiding behind any glamorous surface. Like any muscle that is trained, they develop the capacity to see with more intensity.

It is simply a choice you have to make. At any moment in life you can convert to realism, which is not a belief system at all, but a way of looking at the world. It means every circumstance, every individual is different, and your task is to measure that difference, then take appropriate action. Your eyes are fixed on the world, not on yourself or your ego. What you see determines what you think and how you act. The moment you believe in some cherished idea that you will hold on to no matter what your eyes and ears reveal to you, you are no longer a realist.

To see this power in action, look at a man like Abraham Lincoln, perhaps our greatest president. He had little formal education and grew up in a harsh frontier environment. As a young man, he liked to take apart machines and put them back together. He was practical to the core. As president, he found himself having to confront the gravest crisis in our history. He was surrounded by cabinet members and advisers who were out to promote themselves or some rigid ideology they believed in. They were emotional and heated; they saw Lincoln as weak. He seemed to take a long time to make a decision, and it would often be the opposite of what they had counseled. He trusted generals like Ulysses S. Grant, who was an alcoholic and a social misfit. He worked with those whom his advisers considered political enemies on the other side of the aisle.

What they didn't realize at the time was that Lincoln came to each circumstance without preconceptions. He was determined to measure everything exactly as it was. His choices were made out of pure pragmatism. He was a keen observer of human nature and stuck with Grant because he saw him as the only general capable of effective action. He judged people by results, not friendliness or political values. His careful weighing of people and events was not a weakness but the height of strength, a fearless quality. And working this way, he carefully guided the country past countless dangers. It is not a history we are accustomed to reading about, since we prefer to be swept up in great ideas and dramatic gestures. But the genius of Lincoln was his ability to focus intensely on reality and see things for what they were. He was a living testament to the power of realism.

It might seem that seeing so much of reality could make one depressed, but the opposite is the case. Having clarity about where you are headed, what people are up to, and what is happening in the world around you will translate into confidence and power, a sensation of lightness. You will feel more connected to your environment, like a spider on its web. Whenever things go wrong in life you will be able to right yourself faster than others, because you will quickly see what is really going on and how you can exploit even the worst moment. And once you taste this power, you will find more satisfaction from an intense absorption in reality than from indulging in any kind of fantasy.

Keys to Fearlessness

KNOW THE OTHER, KNOW YOURSELF, AND THE VICTORY WILL NOT BE AT RISK; KNOW THE GROUND, KNOW THE NATURAL CONDITIONS, AND THE VICTORY WILL BE TOTAL.

—Sun Tzu

America was once a country of great realists and pragmatists. This came from the harshness of the environment, the many dangers of frontier life. We had to become keen observers of everything going on around us to survive. In the nineteenth century, such a way of looking at the world led to innumerable inventions, the accumulation of wealth, and the emergence of our country as a great power. But with this growing power, the environment no longer pressed upon us so violently, and our character began to change.

Reality came to be seen as something to avoid. Secretly and slowly we developed a taste for escape—from our problems, from work, from the harshness of life. Our culture began to manufacture endless fantasies for us to consume. And fed on such illusions, we became easier to deceive, since we no longer had a mental barometer for distinguishing fact from fiction.

This is a dynamic that has repeated itself throughout history. Ancient Rome began as a small city-state. Its citizens were tough and stoic. They were famous for their pragmatism. But as they moved from being a republic to an empire and their power expanded, everything reversed itself. Their citizens' minds hungered for newer and newer forms of escape. They lost all sense of proportion—petty political battles consumed their attention more than much larger dangers on the outskirts of the empire. The empire fell well before the invasion of the barbarians. It collapsed from the collective softness of its citizens' minds and the turning of their back on reality.

Understand: as an individual you cannot stop the tide of fantasy and escapism sweeping a culture. But you can stand as an individual bulwark to this trend and create power for yourself. You were born with the greatest weapon in all of nature — the rational, conscious mind. It has the power to expand your vision far and wide, giving you the unique capacity to distinguish patterns in events, learn from the past, glimpse into the future, see through appearances. Circumstances are conspiring to dull that weapon and render it useless by turning you inward and making you afraid of reality.

Consider it war. You must fight this tendency as best you can and move in the opposite direction. You must turn outward and become a keen observer of all that is around you. You are doing battle against all the fantasies that are thrown at you. You are tightening your connection to the environment. You want clarity, not escape and confusion. Moving in this direction will instantly bring you power among so many dreamers.

Regard the following as exercises for your mind—to make it less rigid, more penetrating and expansive, a sharper gauge of reality. Practice all of them as often as you can.

REDISCOVER CURIOSITY—OPENNESS

One day it came to the attention of the ancient Greek philosopher Socrates that the oracle at Delphi had pronounced him the wisest man in the world. This baffled the philosopher—he did not think himself worthy of such a decree. It made him uncomfortable. He decided to simply go around Athens and find a person who was wiser than he—that should be easy and it would disprove the oracle.

He engaged in many discussions with politicians, poets, craftsmen, and fellow philosophers. He began to realize that the oracle was right. All the people he talked to had such a certainty about things, venturing solid opinions about matters of which they had no experience; they were full of so much air. If you questioned them at all, they could not really defend their opinions, which seemed based on something they had decided years earlier. His superiority, he realized, was that he knew that he knew nothing. This left his mind open to experiencing things as they are, the source of all knowledge.

This position of basic ignorance was what you had as a child. You had a need and hunger for knowledge, to overcome this ignorance, so you observed the world as closely as possible, absorbing large amounts of information. Everything was a source of wonder. With time our minds tend to close off. At some point, we feel like we know what we need to know; our opinions are certain and firm. We do this out of fear. We don't want our assumptions about life challenged. If we go too far in this direction, we can become extremely defensive and cover up our fears by acting with supreme confidence and certainty.

What you need to do in life is return to that mind you possessed as a child, opening up to experience instead of closing it off. Just imagine for a day that you do not know anything, that what you believe could be completely false. Let go of your preconceptions and even your most cherished beliefs. Experiment. Force yourself to hold the opposite opinion or see the world through your enemy's eyes. Listen to the people around you with more attentiveness. See everything as a source for education—even the most banal encounters. Imagine that the world is still full of mystery.

When you operate this way, you will notice that something strange often happens. Opportunities will begin to fall into your lap because you are suddenly more receptive to them. Sometimes luck or serendipity is more a function of the openness of your mind.

KNOW THE COMPLETE TERRAIN—EXPANSION

War is fought over specific terrain. But there is more involved than just that. There is also the morale of the enemy soldiers, the political leaders who set them in motion, the minds of the opposing generals who make the key decisions, and the money and resources that stand behind it all. A mediocre general will confine his knowledge to the physical terrain. A better general will try to expand his knowledge by reading reports about the other factors that influence an army. And the superior general will try to intensify this knowledge by observing as much as he can with his own eyes or consulting firsthand sources. Napoleon Bonaparte is the greatest general who ever lived, and what elevated him above all others was the mass of information he absorbed about all of the details of battle, with as few filters as possible. This gave him a superior grasp on reality.

Your goal is to follow the path of Napoleon. You want to take in as much as possible with your own eyes. You communicate with people up and down the chain of command within your organization. You do not draw any barriers to your social interactions. You want to expand your access to different ideas. Force yourself to go to events and places that are beyond your usual circle. If you cannot observe something firsthand, try to get reports that are more direct and less filtered, or vary the sources so that you can see things from several sides. Get a fingertip feel for everything going on in your environment—the complete terrain.

DIG TO THE ROOTS—DEPTH

Malcolm X was a realist—he had a way of looking at the world that was honed by years on the streets and in prison. After prison, his mission in life was to figure out the source of the problem for blacks in America. As he explained in his autobiography, "This country goes in for the surface glossing over, the escape ruse, surfaces, instead of truly dealing with its deep-rooted problems." He decided to dig as far below the surface as possible. Finally he arrived at what he believed to be the root cause—dependency. As it stood, African Americans couldn't do things completely on their own—they depended on the government, on liberals, on their leaders, on everybody but themselves. If they could end this dependency, they would have the power to reverse everything.

Malcolm X died before he could go further with his life's mission, but his method remains valid for all time. When you do not get to the root of a problem, you cannot solve it in any meaningful manner. People like to look at the surfaces, get all emotional and react, doing things that make them feel better in the short term but do nothing for them in the long term.

This must be the power and the direction of your mind whenever you encounter some problem—to bore deeper and deeper until you get at something basic and at the root. Never be satisfied with what presents itself to your eyes. See what underlies it all, absorb it, and then dig deeper. Always question why this particular event has happened, what the motives of the various actors are, who really is in control, who benefits by this action. Often, it will revolve around money and power—that is what people are usually fighting over, despite the surface gloss they give to it. You may never get to the actual root, but the process of digging will bring you closer. And operating in this way will help develop your mind into a powerful analytical instrument.

SEE FURTHER AHEAD—PROPORTION

By our nature as rational, conscious creatures, we cannot help but think of the future. But most people, out of fear, limit their view of the future to a narrow range—thoughts of tomorrow, a few weeks ahead, perhaps a vague plan for the months to come. We are generally dealing with so many immediate battles, it is hard for us to lift our gaze above the moment. It is a law of power, however, that the further and deeper we contemplate the future, the greater our capacity to shape it according to our desires.

If you have a long-term goal for yourself, one that you have imagined in detail, then you are better able to make the proper decisions in the present. You know which battles or positions to avoid because they don't advance you towards your goal. With your gaze lifted to the future, you can focus on the dangers looming on the horizon and take proactive measures to avert them. You have a sense of proportion—sometimes the things we fuss over in the present don't matter in the long run. All of this gives you an increased power to reach your objectives.

As part of this process, look at the smaller problems that are plaguing you or your enterprise in the present, and draw arrows to the future, imagining what they could possibly lead to if they grow larger. Think of your own biggest mistakes or those of others. How could they have been foreseen? Generally there are signs that seem so obvious afterwards. Now imagine those very same signs that you are probably ignoring in the present.

LOOK AT PEOPLE'S DEEDS, NOT WORDS—SHARPNESS

In war or any competitive game, you don't pay attention to people's good or bad intentions. They don't matter. It should be the same in the game of life. Everyone is playing to win, and some people will use moral justifications to advance their side. All you look at are people's maneuvers—their actions in the past and what you might expect in the future. In this area, you are fiercely realistic. You understand that everyone is after power, and that to get it we all occasionally manipulate and even deceive. That is human nature and there is no shame in it. You don't take people's maneuvers personally; you merely try to defend or advance yourself.

As part of this approach, you must become a better observer of people. This cannot be done on the Internet. It must be honed in personal interactions. You are trying to read people, see through them as best you can. You come to understand, for instance, that a person who is too obviously friendly after too short a time is often up to no good. If they flatter you, it is generally out of envy. Behavior that stands out and seems excessive is a sign. Don't get caught up in people's grand gestures, in the public face they put on. Pay more attention to the details, to the little things they reveal in their day-to-day lives. Their decisions reveal a lot, and you can often discern a pattern if you look at them closely.

In general, looking at people through the lens of your emotions will cloud what you see and make you misunderstand everything. What you want is a sharp eye towards your fellow humans—one that is piercing, objective, and nonjudgmental.

REASSESS YOURSELF—DETACHMENT

Your increasing powers of observation must occasionally be aimed at yourself. Think of this as a ritual you will engage in every few weeks—a rigorous reassessment of who you are and where you are headed. Look at your most recent actions as if they were the maneuvers of another person. Imagine how you could have done it all better—avoided unnecessary battles or confronted people who stood in your way, instead of running away from them. The goal here is not to beat up on yourself but to have the capacity to adapt and change your behavior by moving closer to the reality.

The endgame of such an exercise is to cultivate the proper sense of detachment from yourself and from life. It is not that you want to feel this detachment at every moment. There are times that require you to act with heart and boldness, without doubts or self-distance. On many occasions, however, you need to be able to assess what is happening, without your ego or emotions coloring your perceptions. Moving to a calm, detached inner position to observe events will become a habit and something you can rely on amid any crisis. At those moments in life when others lose their balance, you will find yours with relative ease. As a person who cannot be easily ruffled by events, you will attract attention and power.

Reversal of Perspective

The word "realist" often comes with some negative connotations. Realists, according to conventional wisdom, can be practical to a fault; they often lack a feel for the finer, higher things in life. Taken too far, such types can be cynical, manipulative, Machiavellian. They stand in contrast to dreamers, people of high imagination who inspire us with their ideals or divert us with their fantastical creations.

This is a concept that comes from looking at the world through the lens of fear. It is time we reverse this perspective and see dreamers and realists in their true light. The dreamers, those who misread the actual state of affairs and act upon their emotions, are often the source of the greatest mistakes in history—the wars that are not thought out, the disasters that are not foreseen. Realists, on the other hand, are the real inventors and innovators. They are men and women of imagination, but their imagination is in close contact with the environment, with reality—they are empirical scientists, writers with a sharp understanding of human nature, or leaders who guide us thoughtfully through crises. They are strong enough to see the world as it is, including their own personal inadequacies.

Let us take this further. The real poetry and beauty in life comes from an intense relationship with reality in all its aspects. Realism is in fact the ideal we must aspire to, the highest point of human rationality.

PEOPLE WHO CLING TO THEIR DELUSIONS FIND IT DIFFICULT, IF NOT IMPOSSIBLE, TO LEARN ANYTHING WORTH LEARNING: A PEOPLE UNDER THE NECESSITY OF CREATING THEMSELVES MUST EXAMINE EVERYTHING, AND SOAK UP LEARNING THE WAY THE ROOTS OF A TREE SOAK UP WATER.

—James Baldwin

CHAPTER 2

Make Everything Your Own— Self-Reliance

WHEN YOU WORK FOR OTHERS, YOU ARE AT THEIR
MERCY. THEY OWN YOUR WORK; THEY OWN YOU. YOUR
CREATIVE SPIRIT IS SQUASHED. WHAT KEEPS YOU IN
SUCH POSITIONS IS A FEAR OF HAVING TO SINK OR
SWIM ON YOUR OWN. INSTEAD YOU SHOULD HAVE A
GREATER FEAR OF WHAT WILL HAPPEN TO YOU IF YOU
REMAIN DEPENDENT ON OTHERS FOR POWER. YOUR
GOAL IN EVERY MANEUVER IN LIFE MUST BE OWNER-
SHIP, WORKING THE CORNER FOR YOURSELF. WHEN
IT IS YOURS, IT IS YOURS TO LOSE—YOU ARE MORE
MOTIVATED, MORE CREATIVE, MORE ALIVE. THE
ULTIMATE POWER IN LIFE IS TO BE COMPLETELY SELF-
RELIANT, COMPLETELY YOURSELF.

The Hustler's Empire

HUMAN NATURE IS SO CONSTITUTED, THAT IT CANNOT
HONOR A HELPLESS MAN, ALTHOUGH IT CAN PITY HIM;
AND EVEN THIS IT CANNOT DO LONG, IF THE SIGNS OF
POWER DO NOT ARISE.

—Frederick Douglass

After serving a short sentence in a Brooklyn rehabilitation program for his first offense as a drug dealer, Curtis Jackson returned to the streets virtually back at square one. The money
he had earned the previous few years as a corner hustler was
all gone, and his once loyal customers had all found other dealers to buy from.

A friend, now running a fairly large crack-cocaine operation, offered Curtis a job bagging up drugs. He would be paid
a daily wage, and not a bad one. Curtis desperately needed the
money, so he accepted the offer. Perhaps further down the road

his friend would cut him in on some of the action and he could reestablish his own business. But from the first day on the job, he realized that this was all a mistake. He was working with a group of other baggers, all former dealers. They were now hired help; they had to show up at a certain time and bow down to the authority of their employers. Curtis had lost not only his money but also his freedom. This new position went against all of the survival lessons he had learned up till then in his short life.

Curtis had never known his father, and his mother had been murdered when he was eight years old. His grandparents had essentially raised him; they were loving and kind, but they had a lot of children to look after and not much time to give individual attention. If he wanted any kind of guidance or advice, there was nobody in his life to turn to. At the same time, if he wanted anything new, such as clothes, he did not feel comfortable asking his grandparents—they did not have much money. What all of this meant was that he was essentially alone in this world. He could not rely on anyone to give him anything. He would have to fend for himself.

Then crack cocaine exploded on the streets in the mid-1980s and everything changed in neighborhoods like his. In the past, large gangs controlled the drug business, and to be involved you had to fit into their structure and spend years moving up the ladder. But crack was so easy to manufacture and the demand was so high, that anyone—no matter how young—could get in on the game without any startup capital. You could work on your own and make good money. For those like Curtis who grew up with little parental supervision and a disdain for authority, being a corner dealer was the perfect fit—no political games, no bosses above you. And so he quickly joined the growing pool of hustlers dealing crack on the streets of Southside Queens.

As he got further into the game, he learned a fundamental lesson. There were endless problems and dangers confronting the street hustler—undercover cops, fiends, and rival dealers scheming to rob you. If you were weak, you looked for others to help you or for some crutch to lean on, such as drugs or alcohol. This was the path of doom. Eventually your friend would not show up as promised or your mind would be too clouded by drugs to see someone's treachery. The only way to survive was to admit you were on your own, learn to make your own decisions, and trust your judgment. Do not ask for what you need but take it. Depend only on your wits.

It was as if a hustler, born amid squalor and cramped quarters, possessed an empire. This was not something physical—the corner that he worked or the neighborhood he wanted to take over. It was his time, his energy, his creative schemes, his freedom to move where he wanted to. If he kept command of that empire, he would make money and thrive. If he looked for help, if he got caught up in other people's political games, he gave all of that away. In such a case, the negative conditions of the hood would be magnified and he would end up a beggar, a pawn in someone else's game.

As he sat there bagging drugs that first day, Curtis realized that this went far beyond a momentary lull in his life in which he needed some quick money. This was a turning point. He looked at the other baggers. They all had suffered downturns in fortune—violence, prison time, etc. They had become scared and tired of the grind. They wanted the comfort and security of a paycheck. And this would become the pattern for the rest of their lives—afraid of life's challenges, they would come to depend on other people to help them. Perhaps they could go on like this for several years, but the day of reckoning would come when there were no more jobs and they had forgotten how to fend for themselves.

It was ludicrous for Curtis to imagine that the man now employing him to bag would some day help him set up shop. Bosses don't do things like that, even if they're your friends. They think of themselves and they use you. He had to get out now, before that empire slipped from his hands and he became yet another former hustler dependent on favors.

He quickly went into full hustling mode and figured his way out of the trap. At the end of the first day, he made a deal with the baggers. He would dole out the daily cash he had been paid for the job to all of them. In return, he would teach them how to put less crack in each capsule but make it look full (he had been doing this on the street for years). They were then to give Curtis the extra crack that was left over from each capsule. Within a week, he had accumulated enough drugs to return to hustling on the streets, on his terms. After that, he swore to himself he'd never work for another person ever again. He would rather die.

Years later, Curtis (now known as 50 Cent) had managed to segue into a music career, and after a fierce mix-tape campaign on the streets of New York in which he became a local celebrity, he gained the attention of Eminem, who helped sign him to a lucrative deal on his own label within Interscope Records.

For the launch of his debut album, *Get Rich or Die Tryin'*, there was a lot of work to do—a marketing campaign, videos, artwork—and so he went to Los Angeles to work with Interscope on these projects. But the more time he spent in their cushy offices, the more he had the feeling that he was at yet another turning point in his life.

The game these music executives were playing was simple: They owned your music and a lot more. They wanted to package the artist in their way, and this dictated all of the key decisions on the music videos and publicity. In return, they lavished you with money and perks. They created a feeling of dependence—without their massive machine behind you, you were helpless in the face of a viciously competitive business. In essence, you were exchanging money for freedom. And once you internally succumbed to their logic and their money, you were finished. You were a high-paid bagger doing a job.

And so, as before, Fifty went into full hustling mode to reclaim his empire. In the short term, he schemed to shoot his own videos, with his own money, and come up with his own marketing schemes. To Interscope it seemed like he was saving them time and resources, but to Fifty it was a subtle way to regain control over his image. He set up a record label for his own stable of artists from within Interscope and he used this label to teach himself all aspects of production. He created his own website where he could experiment with new ways to market his music. He turned the dependence dynamic around, using Interscope as a school for teaching him how to run things on his own.

All of this was part of the endgame he had in mind—he would run out his contract with Interscope, and instead of re-negotiating a new one, he would proclaim his independence and be the first artist to set up his own freestanding record label. From such a position of power, he would have no more executives to please and he could expand his empire on his own terms. It would be just like the freedom he had experienced on the streets, but on a global scale.

The Fearless Approach

I WAS BORN ALONE AND I WILL DIE ALONE. I'VE GOT TO DO WHAT'S RIGHT FOR ME AND NOT LIVE MY LIFE THE WAY ANYBODY ELSE WANTS IT.

—50 Cent

You came into this life with the only real possessions that ever matter—your body, the time that you have to live, your energy, the thoughts and ideas unique to you, and your autonomy. But over the years you tend to give all of this away. You spend years working for others—they own you during that period. You get needlessly caught up in people's games and battles, wasting energy and time that you will never get back. You come to respect your own ideas less and less, listening to experts, conforming to conventional opinions. Without realizing it you squander your independence, everything that makes you a creative individual.

Before it is too late, you must reassess your entire concept of ownership. It is not about possessing things or money or titles. You can have all of that in abundance but if you are someone who still looks to others for help and guidance, if you depend on your money or resources, then you will eventually lose what you have when people let you down, adversity strikes, or you reach for some foolish scheme out of impatience. True ownership can come only from within. It comes from a disdain for anything or anybody that impinges upon your mobility, from a confidence in your own decisions, and from the use of your time in constant pursuit of education and improvement.

Only from this inner position of strength and self-reliance will you be able to truly work for yourself and never turn back. If situations arise in which you must take in partners or fit within another organization, you are mentally preparing yourself for the moment when you will move beyond these momentary entanglements. If you do not own yourself first, you will continually be at the mercy of people and circumstance, looking outward instead of relying on yourself and your wits.

Understand: we are living through an entrepreneurial revolution, comparable to the one that swept through Fifty's neighborhood in the 1980s, but on a global scale. The old power centers are breaking up. Individuals everywhere want more control over their destiny and have much less respect for an authority that is not based on merit but on mere power. We have all naturally come to question why someone should rule over us, why our source of information should depend on the mainstream media, and on and on. We do not accept what we accepted in the past.

Where we are naturally headed with all of this is the right and capacity to run our own enterprise, in whatever shape or form, to experience that freedom. We are all corner hustlers in a new economic environment and to thrive in it we must cultivate the kind of self-reliance that helped push Fifty past all of the dangerous dependencies that threatened him along the way.

For Fifty it was very clear—he was alone in the house he grew up in and on the streets. He lacked the usual supports and so he was forced to become self-sufficient. The consequences of being dependent on people were so much more severe in his case—it would mean constant disappointment and urgent needs that went unmet. It is harder for us to realize that we are essentially alone in this world and in need of the skills that Fifty had to develop for himself on the streets. We have layers of support that seem to prop us up. But these supports are illusions in the end.

Everyone in the world is governed by self-interest. People naturally think first of themselves and their agendas. An occasional affectionate or helpful gesture from people you know tends to cloud this reality and make you expect more of this support—until you are disappointed, again and again. You are more alone than you imagine. This should not be a source of fear but of freedom. When you prove to yourself that you can get things on your own, then you experience a sense of liberation. You are no longer waiting for people to do this or that for you (a frustrating and infuriating experience). You have confidence that you can manage any adverse situation on your own.

Look at a man like Rubin "Hurricane" Carter—a successful middleweight boxer who found himself arrested in 1966 at the height of his career and charged with a triple murder. The following year he was convicted and sentenced to three consecutive life terms. Through it all Carter vehemently maintained his innocence, and in 1986 he was finally exonerated of the crimes and set free. But for those nineteen years, he had to endure one of the most brutal environments known to man, one designed to break down every last vestige of autonomy.

Carter knew he would be freed at some point. But on the day of his release, would he walk the streets with a spirit crushed by years in prison? Would he be the kind of former prisoner who keeps coming back into the system because he can no longer do anything for himself?

He decided that he would defeat the system—he would use the years in prison to develop his self-reliance so that when he was freed it would mean something. For this purpose he devised the following strategy: He would act like a free man while surrounded by walls. He would not wear their uniform or carry an ID badge. He was an individual, not a number. He would not eat with the other prisoners, do the assigned tasks, or go to his parole hearings. He was placed in solitary confinement for these transgressions but he was not afraid of the punishments, nor of being alone. He was afraid only of losing his dignity and sense of ownership.

As part of this strategy, he refused to have the usual entertainments in his cell—television, radio, pornographic magazines. He knew he would grow dependent on these weak pleasures and this would give the wardens something to take away from him. Also, such diversions were merely attempts to kill time. Instead he became a voracious reader of books that would help toughen his mind. He wrote an autobiography that gained sympathy for his cause. He taught himself law, determined to get his conviction overturned by himself. He tutored other prisoners in the ideas that he had learned through his reading. In this way, he reclaimed the dead time of prison for his own purposes.

When he was eventually freed, he refused to take civil action against the state—that would acknowledge he had been in prison and needed compensation. He needed nothing. He was now a free man with the essential skills to get power in the world. After prison he became a successful advocate for prisoners' rights and was awarded several honorary law degrees.

Think of it this way: dependency is a habit that is so easy to acquire. We live in a culture that offers you all kinds of crutches—experts to turn to, drugs to cure any psychological unease, mild pleasures to help pass or kill time, jobs to keep you just above water. It is hard to resist. But once you give in, it is like a prison you enter that you cannot ever leave. You continually look outward for help and this severely limits your options and maneuverability. When the time comes, as it inevitably does, when you must make an important decision, you have nothing inside of yourself to depend on.

Before it is too late, you must move in the opposite direction. You cannot get this requisite inner strength from books or a guru or pills of any kind. It can come only from you. It is a kind of exercise you must practice on a daily basis—weaning yourself from dependencies, listening less to others' voices and more to your own, cultivating new skills. As happened with Carter and with Fifty, you will find that self-reliance becomes the habit and that anything that smacks of depending on others will horrify you.

‖‖

Keys to Fearlessness

‖‖

I AM OWNER OF MY MIGHT, AND I AM SO WHEN I KNOW MYSELF AS UNIQUE.

—Max Stirner

As children we all faced a similar dilemma. We began life as willful creatures who had yet to be tamed. We wanted and demanded things for ourselves, and we knew how to get them from the adults around us. And yet at the same time, we were completely dependent on our parents for so many important things—comfort, protection, love, guidance. And so from deep inside, we developed an ambivalence. We wanted the freedom and power to move on our own, but we also craved the comfort and security only others could give us.

In adolescence we rebelled against the dependent part of our character. We wanted to differentiate ourselves from our parents and show that we could fend for ourselves. We struggled to form our own identity and not simply conform to our parents' values. But as we get older, that childhood ambivalence tends to return to the surface. In the face of so many difficulties and competition in the adult world, a part of us yearns to return to that childish position of dependence. We maintain an adult face and work to gain power for ourselves, but deep inside we secretly wish that our spouses, partners, friends, or bosses could take care of us and solve our problems.

We must wage a ferocious war against this deeply embedded ambivalence, with a clear understanding of what is at stake. Our task as an adult is to take full possession of that autonomy and individuality we were born with. It is to finally overcome the dependent phase in childhood and stand on our own. We must see the desire for a return to that phase as regressive and dangerous. It comes from fear—of being responsible for our success and failure, of having to act on our own and make the hard decisions. We will often package this as the opposite—that by working for others, being dutiful, fitting in, or subsuming our personality to the group, we are being a good person. But that is our fear speaking and deluding us. If we give in to this fear, then we will spend our lives looking outward for salvation and never find it. We will merely move from one dependency to another.

For most of us, the critical terrain in this war is the work world. Most of us enter adult life with great ambitions for how we will start our own ventures, but the harshness of life wears us down. We settle into some job and slowly give in to the illusion that our bosses care about us and our future, that they spend time thinking of our welfare. We forget the essential truth that all humans are governed by self-interest. Our bosses keep us around out of need, not affection. They will get rid of us the moment that need is less acute or they find someone younger and less expensive to replace us. If we succumb to the illusion and the comfort of a paycheck, we then neglect to build up self-reliant skills and merely postpone the day of reckoning when we are forced to fend for ourselves.

Your life must be a progression towards ownership—first mentally of your independence, and then physically of your work, owning what you produce. Think of the following steps as a kind of blueprint for how to move in this direction.

STEP ONE: RECLAIM DEAD TIME

When Cornelius Vanderbilt (1794–1877) was twelve years old he was forced to work for his father in his small shipping business. It was drudge work and he hated it. Cornelius was a willful, ambitious child, and so in his mind he made the following determination: within a couple of years he was going to start his own shipping enterprise. This simple decision altered everything. Now this job was an urgent apprenticeship. He had to keep his eyes open, learn everything he could about his father's business, including how he could do things better. Instead of dull labor, it was now an exciting challenge.

At the age of sixteen he borrowed $100 from his mother. He used the money to buy a boat and began ferrying passengers between Manhattan and Staten Island. Within a year he paid back the loan. By the time he was twenty-one he had made a small fortune and was on his way to becoming the wealthiest man of his time. From this experience he established his life-long motto: "Never be a minion, always be an owner."

Time is the critical factor in our lives, our most precious resource. The problem when we work for others is that so much of this becomes dead time that we want to pass as quickly as possible, time that is not our own. Almost all of us must begin our careers working for others, but it is always within our power to transform this time from something dead to something alive. If we make the same determination as Vanderbilt—to be an owner and not a minion—then that time is used to learn as much as we can about what is going on around us—the political games, the nuts and bolts of this particular venture, the larger game going on in the business world, how we could do things better. We have to pay attention and absorb as much information as possible. This helps us endure work that does not seem so rewarding. In this way, we own our time and our ideas before owning a business.

Remember: your bosses prefer to keep you in dependent positions. It is in their interest that you do not become self-reliant, and so they will tend to hoard information. You must secretly work against this and seize this information for yourself.

STEP TWO: CREATE LITTLE EMPIRES

While still working for others, your goal at some point must be to carve out little areas that you can operate on your own, cultivating entrepreneurial skills. This could mean offering to take over projects that others have left undone or proposing to put into action some new idea of your own, but nothing too grandiose to raise suspicion. What you are doing is cultivating a taste for doing things yourself—making your own decisions, learning from your own mistakes. If your bosses do not allow you to make such a move on any scale, then you are not in the right place. If you fail in this venture, then you have gained a valuable education. But generally taking on such things on your own initiative forces you to work harder and better. You are more creative and motivated because there is more at stake; you rise to the challenge.

Keep in mind the following: what you really value in life is ownership, not money. If ever there is a choice—more money or more responsibility—you must always opt for the latter. A lower-paying position that offers more room to make decisions and carve out little empires is infinitely preferable to something that pays well but constricts your movements.

STEP THREE: MOVE HIGHER UP THE FOOD CHAIN

In 1499, Pope Alexander VI managed to carve out a principality for his son, Cesare Borgia, in the Romagna district of Italy.

This was not easy. All kinds of rival powers were competing for control of the country—families that dominated the political scene, foreign kings scheming to take over certain regions, city-states with spheres of influence, and finally the church itself. To secure Romagna for his son, the pope had to win over one of the two most powerful families in Italy, make an alliance with King Louis XII of France, and hire a mercenary army.

Cesare Borgia was a shrewd young man. His goal was to expand beyond Romagna and eventually unify all of Italy, making it a great power. But his position now depended on various outside forces that controlled his destiny, each one above the other—the army beholden to the powerful families and king of France, then the pope himself who could die any day and be replaced by someone antagonistic to Borgia. These alliances could shift and turn against him. He had to eliminate these dependencies, one by one, until he could stand on his own, with nobody above him.

Using bribery, he put himself at the head of the family faction his father had allied him with, then moved to eliminate its main rival. He worked to get rid of the mercenary army and establish his own. He schemed to make alliances that would secure him against the French king who now saw him as a threat. He gobbled up more and more regions. He was on the verge of expanding his base to a point of no return when he suddenly fell gravely ill in 1504. Shortly thereafter, his father died and was soon replaced by a pope determined to stop Cesare Borgia. Who knows how far he could have gotten if his plans had not become unraveled by such unforeseen circumstances.

Borgia was a kind of self-reliant entrepreneur before his time. He understood that people are political creatures, continually scheming to secure their own interests. If you form partnerships with them or depend upon them for your advancement and protection, you are asking for trouble. They will either turn against you at some point or use you as a cat's-paw to get what they want. Your goal in life must be to always move higher and higher up the food chain, where you alone control the direction of your enterprise and depend on no one. Since this goal is a future ideal, in the present you must strive to keep yourself free of unnecessary entanglements and alliances. And if you cannot avoid having partners, make sure that you are clear as to what function they serve for you and how you will free yourself of them at the right moment

You must remember that when people give you things or do you favors it is always with strings attached. They want something from you in return—assistance, unquestioned loyalty, and so forth. You want to keep yourself free of as many of these obligations as possible, so get in the habit of taking what you need for yourself instead of expecting others to give it to you.

STEP FOUR: MAKE YOUR ENTERPRISE
A REFLECTION OF YOUR INDIVIDUALITY

Your whole life has been an education in developing the skills and self-reliance necessary for creating your own venture, being your own boss. But there is one last impediment to making this work. Your tendency will be to look at what other people have done in your field, how you could possibly repeat or emulate their success. You can gain some power with such a strategy, but it won't go far and it won't last.

Understand: you are one of a kind. Your character traits are a kind of chemical mix that will never be repeated in history. There are ideas unique to you, a specific rhythm and perspective that are your strengths, not your weaknesses. You must not be afraid of your uniqueness and you must care less and less what people think of you.

This has been the path of the most powerful people in history. Throughout his life the great jazz musician Miles Davis was always being pushed into making his sound fit the particular rage of the time. Instead he kept insisting on putting his own stamp on anything he played. As he got older this became more and more extreme until he revolutionized the jazz world with his constant innovations in sound. At a certain point he simply stopped listening to others. John F. Kennedy refused to run a campaign like Franklin Delano Roosevelt or any other American politician in the past. He created his own inimitable style, based on the times he lived in and his own personality. By going his own way, he forever altered the course of political campaigning.

This uniqueness that you express is not anything wild or too strange. That is an affectation in itself. People are rarely that different. Rather you are being yourself, as far as you can take that. The world cannot help but respond to such authenticity.

Reversal of Perspective

We might think of people who are independent and used to being alone as reclusive, prickly, and hard to be around. In our culture we tend to elevate those who are smooth talkers, seem more gregarious, and fit in better, conforming to certain norms. They smile and seem happier. This is a superficial appraisal of character; if we reverse our perspective and look at this from the fearless point of view we come to the opposite conclusion.

People who are self-sufficient are generally types who are more comfortable with themselves. They do not look for things that they need from other people. Paradoxically this makes them more attractive and seductive. We wish we could be more like that and want to be around them, hoping that some of their independence might rub off on us. The needy, clingy types—often the most sociable—unconsciously push us away. We feel their need for comfort and validation and secretly we want to say to them: "Get it for yourself—stop being so weak and dependent."

Those who are self-reliant turn to people out of strength—a desire for pleasant company or an exchange of ideas. If people do not do what they want or expect, they are not hurt or let down. Their happiness comes from within and is all the more profound for that reason.

Finally, do not be taken in by the culture of ease. Self-help books and experts will try to convince you that you can have what you want by following a few simple steps. Things that come easy and fast will leave you just as fast. The only way to gain self-reliance or any power is through great effort and practice. And this effort should not be seen as something ugly or dull; it is the process of gaining power over yourself that is the most satisfying of all, knowing that step-by-step you are elevating yourself above the dependent masses.

THERE IS A TIME IN EVERY MAN'S EDUCATION WHEN HE ARRIVES AT THE CONVICTION THAT . . . IMITATION IS SUICIDE . . . THAT THOUGH THE WIDE UNIVERSE IS FULL OF GOOD, NO KERNEL OF NOURISHING CORN CAN COME TO HIM BUT THROUGH HIS TOIL BESTOWED ON THAT PLOT OF GROUND WHICH IS GIVEN TO HIM TO TILL. THE POWER WHICH RESIDES IN HIM IS NEW IN NATURE, AND NONE BUT HE KNOWS WHAT THAT IS WHICH HE CAN DO, NOR DOES HE KNOW UNTIL HE HAS TRIED.

—Ralph Waldo Emerson

CHAPTER 3

Turn Shit into Sugar— Opportunism

EVERY NEGATIVE SITUATION CONTAINS THE POSSIBIL-
ITY FOR SOMETHING POSITIVE, AN OPPORTUNITY. IT
IS HOW YOU LOOK AT IT THAT MATTERS. YOUR LACK OF
RESOURCES CAN BE AN ADVANTAGE, FORCING YOU TO
BE MORE INVENTIVE WITH THE LITTLE THAT YOU HAVE.
LOSING A BATTLE CAN ALLOW YOU TO FRAME YOURSELF
AS THE SYMPATHETIC UNDERDOG. DO NOT LET FEARS
MAKE YOU WAIT FOR A BETTER MOMENT OR BECOME
CONSERVATIVE. IF THERE ARE CIRCUMSTANCES YOU
CANNOT CONTROL, MAKE THE BEST OF THEM. IT IS THE
ULTIMATE ALCHEMY TO TRANSFORM ALL SUCH NEGA-
TIVES INTO ADVANTAGES AND POWER.

Hood Alchemy

IF ONE IS CONTINUALLY SURVIVING THE WORST THAT
LIFE CAN BRING, ONE EVENTUALLY CEASES TO BE CON
TROLLED BY A FEAR OF WHAT LIFE CAN BRING.

—James Baldwin

For well over a year 50 Cent had been working on what was meant to be his debut album, *Power of the Dollar*, and finally in the spring of 2000 it was ready to be released by Columbia Records. It represented to him all the struggles he had been through on the streets, and he had hopes that it would turn his life around for good. In May of that year, however, a few weeks before the launch date, a hired assassin shot nine bullets into him while he sat in the back of a car, one bullet going through his jaw and nearly killing him.

In a flash, all of the momentum he had built up reversed itself. Columbia canceled the release of the record and dropped

Fifty from his contract. There was too much violence associated with him; it was bad for business. A few inquiries made it clear that other labels felt the same—he was being blackballed from the industry. One executive told him flatly he would have to wait at least two years before he could think of resurrecting his career.

The assassination attempt was the result of an old drug beef from his days as a dealer; the killers could not afford to let him survive and would try to finish the job. Fifty had to keep a low profile. At the same time, he had no money and could not return to street hustling. Even many of his friends, who had hoped to be part of his success as a rapper, started to avoid him.

In just a few short weeks he had gone from being poised for fame and fortune to hitting the bottom. And there seemed no way to move out of the corner he found himself in. Could this be the end of all his efforts? It would have been better to die that day than to feel this powerlessness. But as he lay in bed at his grandparents' house, recovering from the wounds, he listened a lot to the radio, and what he heard gave him an incredible rush of optimism: an idea started taking shape in his mind that the shooting was in fact a great blessing in disguise, that he had narrowly survived for a reason.

The music on the radio was all so packaged and produced. Even the tough stuff, the gangsta rap, was fake. The lyrics did not reflect anything from the streets that he knew. The attempt to pass it off as real and urban angered him to a point he could not endure. This was not the time for him to be afraid and depressed, or to sit around and wait a few years while all of the violence around him died down. He had never been a fake studio gangsta and now he had the nine bullet wounds to prove it. This was the moment to convert all of his anger and dark emotions into a powerful campaign that would shake the very foundations of hip-hop.

As a hustler on the streets Fifty had learned a fundamental lesson: Access to money and resources is severely limited in the hood. A hustler must transform every little event and every trifling object into some gimmick for making money. Even the worst shit that happens to you can be converted into gold if you are clever enough. All of the negative factors now facing him—little money, no connections, the price on his head—could be turned into their opposites, advantages and opportunities. That is how he would confront the seemingly insurmountable obstacles now in his path.

He decided to disappear for a few months and, holed up in various friends' houses, he began to re-create himself and his music career. With no executives to have to please or worry about, he could push his lyrics and the hard sounds as far as he wanted. His voice had changed as a result of the pieces of bullet still lodged in his tongue—it now had a hiss. It was still painful for him to move his mouth, so he had to rap more slowly. Instead of trying to normalize and retrain his voice, he determined to turn it into a virtue. His new style of rapping would be more deliberate and menacing; that hiss would remind listeners of the bullet that had gone through his jaw. He would play all of that up.

In the summer of 2001, just as people had begun to forget about him, Fifty suddenly released his first song to the streets. It was called "Fuck You," the title and the lyrics summarizing how he felt about his killers—and everyone who wanted him to go away. Just putting out the song was message enough—he was defying his assassins openly and publicly. Fifty was back, and to shut him up they would have to finish the job. The palpable anger in his voice and the hard-driving sound of the song made it a sensation on the streets. It also came with an added punch—because he seemed to be inviting more violence, the public had to grab up everything he produced before he was killed. The life-and-death angle made for a compelling spectacle.

Now the songs started to pour out of him. He fed off all the anger he felt and the doubts people had had about him. He was also consumed with a sense of urgency—this was his last chance to make it and so he worked night and day. Fifty's mixtapes began to hit the street at a furious pace.

Soon he realized the greatest advantage he possessed in this campaign—the feeling that he had already hit bottom and had nothing to lose. He could attack the record industry and poke fun at its timidity. He could pirate the most popular songs on the radio and put his own lyrics over them to create wicked parodies. He didn't care about the consequences. And the further he took this the more his audiences responded. They loved the transgressive edge to it. It was like a crusade against all the fake crap on the radio, and to listen to Fifty was to participate in the cause.

On and on he went, transforming every conceivable negative into a positive. To compensate for the lack of money to distribute his mix-tapes far and wide, he decided to encourage bootleggers to pirate his tracks and spread his music around like a virus. With the price still on his head, he could not give concerts or do any kind of public promotion; but somehow he turned even this into a marketing device. Hearing his music everywhere but not being able to see him only added to the mystique and the attention people paid to him. Rumors and word of mouth helped form a kind of Fifty mythology. He made himself even scarcer to feed this process.

The momentum now was devastating—you could not go far in New York without hearing his music blasted from some corner. Soon one of his mix-tapes reached the ears of Eminem, who decided this was the future of hip-hop and quickly signed Fifty in early 2003 to his and Dr. Dre's label, Shady Aftermath, completing one of the most rapid and remarkable turnarounds in fortune in modern times.

||||||||||||||||||||||||||||||||||||||||||||||||||||||||||||||||||||||||||||||||||||||||||||||||||||||||||||||||||||||||||||

The Fearless Approach

||||||||||||||||||||||||||||||||||||||||||||||||||||||||||||||||||||||||||||||||||||||||||||||||||||||||||||||||||||||||||||

EVERY NEGATIVE IS A POSITIVE. THE BAD THINGS THAT
HAPPEN TO ME, I SOMEHOW MAKE THEM GOOD. THAT
MEANS YOU CAN'T DO ANYTHING TO HURT ME.

—50 Cent

Events in life are not negative or positive. They are completely neutral. The universe does not care about your fate; it is indifferent to the violence that may hit you or to death itself. Things merely happen to you. It is your mind that chooses to interpret them as negative or positive. And because you have layers of fear that dwell deep within you, your natural tendency is to interpret temporary obstacles in your path as something larger—setbacks and crises.

In such a frame of mind, you exaggerate the dangers. If someone attacks and harms you in some way, you focus on the money or position you have lost in the battle, the negative publicity, or the harsh emotions that have been churned up. This causes you to grow cautious, to retreat, hoping to spare yourself more of these negative things. It is a time, you tell yourself, to lay low and wait for things to get better; you need calmness and security.

What you do not realize is that you are inadvertently making the situation worse. Your rival only gets stronger as you sit back; the negative publicity becomes firmly associated with you. Being conservative turns into a habit that carries over into less difficult moments. It becomes harder and harder to move to the offensive. In essence you have chosen to cast life's inevitable twists of fortune as hardships, giving them a weight and endurance they do not deserve.

What you need to do, as Fifty discovered, is take the opposite approach. Instead of becoming discouraged and depressed by any kind of downturn, you must see this as a wake-up call, a challenge that you will transform into an opportunity for power. Your energy levels rise. You move to the attack, surprising your enemies with boldness. You care less what people think about you and this paradoxically causes them to admire you—the negative publicity is turned around. You do not wait for things to get better—you seize this chance to prove yourself. Mentally framing a negative event as a blessing in disguise makes it easier for you to move forward. It is a kind of mental alchemy, transforming shit into sugar.

Understand: we live in a society of relative prosperity, but in many ways this turns out to be a detriment to our spirit. We come to feel that we naturally deserve good things, that we have certain privileges due to us. When setbacks occur, it is almost a personal affront or punishment. "How could this have happened?" we ask. We either blame other people or we blame ourselves. In both cases, we lose valuable time and become unnecessarily emotional.

In places like the hood or in any kind of materially impoverished environment, the response to hardship is much different. There, bad things happening assume a kind of normality. They are part of daily life. The hustler thinks: "I must make the most of what I have, even the bad stuff, because things are not going to get better on their own. It is foolish to wait; tomorrow may bring even worse shit." If Fifty had waited, as he had been counseled, he would be just another rapper who had had a moment of success and then faded quickly away. The hood would have consumed him.

This hustler mind-set is more realistic and effective. The truth is that life is by nature harsh and competitive. No matter how much money or resources you have accumulated, someone will try to take them from you, or unexpected changes in the world will push you backward. These are not adverse circumstances but merely life as it is. You have no time to lose to fear and depression, and you do not have the luxury of waiting.

All of the most powerful people in history demonstrate in one way or another this fearless attitude towards adversity. Look at George Washington. He was a wealthy landowner but his attitude towards life had been forged by years fighting for the British in the French and Indian War, amid the harsh environment of frontier America. In 1776, Washington was made supreme commander of the American Revolutionary army. At first glance this position seemed more like a curse. The army was a semi-organized mob. It had no training, was poorly paid and outfitted, and its morale was low—most of the soldiers did not really believe they could succeed in defeating the all-powerful British.

Throughout 1777, British forces pushed this weak American army around, from Boston to New York, until by the end of the year Washington had been forced to retreat to New Jersey. This was the darkest moment in his career and in the war for independence. Washington's army had dwindled to a few thousand men; they had little food and were poorly clothed, during one of the bitterest winters in memory. The American Continental Congress, fearing imminent disaster, fled from Philadelphia to Baltimore.

Assessing this situation, a cautious leader would have chosen to wait out the winter, muster more troops, and hope for some change in fortune. But Washington had a different mind-set. As he perceived it, his army would be considered by the British as too weak to pose any threat. Being small, his army could move without the enemy's knowledge, and launch an attack that was all the more surprising for coming out of nowhere. Moving to the attack would excite the troops and gain some much-needed positive publicity. Thinking in this manner, he decided to lead a raid on an enemy garrison in Trenton, which proved to be a great success. He followed this up with an attack on British supplies at Princeton. These daring victories captivated the American public. Confidence had been restored in Washington as a leader and the American army as a legitimate force.

From then on, Washington waged a guerrilla-style war, wearing out the British with the great distances they had to cover. Everything was turned around—lack of funds and experience led to a more creative way of fighting. The smallness of his forces allowed him to torment the enemy with fluid maneuvering over rough terrain. At no point did he decide to wait for more troops or more money or better circumstances—he went continually on the attack with what he had. It was a campaign of supreme fearlessness, in which all negatives were converted into advantages.

This is a common occurrence in history: almost all great military and political triumphs are preceded by some kind of crisis. That is because a substantial victory can only come out of a moment of danger and attack. Without these moments, leaders are never challenged, never get to prove themselves. If the path is too smooth, they grow arrogant and make a fatal mistake. The fearless types require some kind of adversity against which they can measure themselves. The tenseness of such dark moments brings out their creativity and urgency, making them rise to the occasion and turn the tide of fortune from defeat to a great victory.

You must adopt an attitude that is the opposite to how most people think and operate. When things are going well, that is precisely when you must be concerned and vigilant. You know it will not last and you will not be caught unprepared. When things are going badly, that is when you are most encouraged and fearless. Finally you have material for a powerful reversal, a chance to prove yourself. It is only out of danger and difficulty that you can rise at all. By simply embracing the moment as something positive and necessary you have already converted it into gold.

Keys to Fearlessness

[I]N NOOKS ALL OVER THE EARTH SIT MEN WHO ARE WAITING, SCARCELY KNOWING IN WHAT WAY THEY ARE WAITING, MUCH LESS THAT THEY ARE WAITING IN VAIN. OCCASIONALLY THE CALL THAT AWAKENS— THAT ACCIDENT WHICH GIVES THE "PERMISSION" TO ACT—COMES TOO LATE, WHEN THE BEST YOUTH AND STRENGTH FOR ACTION HAS ALREADY BEEN USED UP BY SITTING STILL; AND MANY HAVE FOUND TO THEIR HORROR WHEN THEY "LEAPED UP" THAT THEIR LIMBS HAD GONE TO SLEEP AND THEIR SPIRIT HAD BECOME TOO HEAVY. "IT IS TOO LATE," THEY SAID TO THEM-

SELVES, HAVING LOST THEIR FAITH IN THEMSELVES
AND HENCEFORTH FOREVER USELESS.

—Friedrich Nietzsche

Our minds possess powers we have not even begun to tap into. These powers come from a mix of heightened concentration, energy, and ingenuity in the face of obstacles. Each of us has the capacity to develop these powers, but first we have to be aware of their existence. This is difficult, however, in a culture that emphasizes material means—technology, money, connections—as the answer to everything. We place unnecessary limits on what the mind can accomplish, and that becomes our reality. Look at our concept of opportunity and you will see this in its clearest light.

According to conventional wisdom, an opportunity is something that exists out there in the world; if it comes our way and we seize it, it brings us money and power. This could be a particular job, the perfect fit for us; it could be a chance to create or join a new venture. It could be meeting the appropriate person. In any event, it depends on being at the right place at the right time and having the proper skills to take advantage of this propitious moment. We generally believe there are only a few such golden chances in life, and most of us are waiting for them to cross our path.

This concept is extremely limited in scope. It makes us dependent on outside forces. It stems from a fearful, passive attitude towards life that is counterproductive. It constrains our minds to a small circle of possibility. The truth is that for the human mind, everything that crosses its path can be a potential tool for power and expansion.

Many of us have had the following experience: we find ourselves in an urgent, difficult situation. Perhaps we have to get something done in an impossibly short amount of time, or someone we had counted on for help does not come through, or we are in a foreign land and must suddenly fend for ourselves. In these situations, necessity crowds in on us. We have to get work done and figure out problems quickly or we suffer immediate consequences. What usually happens is that our minds snap to attention. We find the necessary energy because we have to. We pay attention to details that normally elude us, because they might spell the difference between success and failure, life and death. We are surprised at how inventive we become. It is at such moments that we get a glimpse of that potential mental power within us that generally lies untapped. If only we could have such a spirit and attitude in everyday life.

This attitude is what we shall call "opportunism." True opportunists do not require urgent, stressful circumstances to become alert and inventive. They operate this way on a daily basis. They channel their aggressive energy into hunting down possibilities for expansion in the most banal and insignificant events. Everything is an instrument in their hands, and with this enlarged notion of opportunity, they create more of it in their lives and gain great power.

Perhaps the greatest opportunist in history is Napoleon Bonaparte. Nothing escaped his attention. He focused with supreme intensity on all of the details, finding ways to transform even the most trivial aspects of warfare—how to march and carry supplies, how to organize troops into divisions—into tools of power. He ruthlessly exploited the slightest mistake of his opponents. He was the master at turning the worst moments in battle into material for a devastating counterattack.

All of this came out of Napoleon's determination to see everything around him as an opportunity. By looking for these opportunities, he found them. This became a mental skill that he refined to an art. This power is open to each and every one of us if we put into practice the following four principles of the art.

MAKE THE MOST OF WHAT YOU HAVE

In 1704, a Scottish sailor named Alexander Selkirk found himself marooned on a deserted island some four hundred miles off the coast of Chile. All he had with him was a rifle, some gunpowder, a knife, and some carpenter's tools. In exploring the interior, he saw nothing but a bunch of goats, cats, rats, and some unfamiliar animals that made strange noises at night. It was a shelterless environment. He decided to keep to the shoreline, slept in a cave, found enough to eat by catching fish, and slowly gave way to a deep depression. He knew he would run out of gunpowder, his knife would get rusty, and his clothes would rot on his back. He could not survive on just fish. He did not have enough supplies to get by and the loneliness was crushing. If only he had brought over more materials from his ship.

Then suddenly the shoreline was invaded by sea lions; it

was their mating season. Now he was forced to move inland. There, he could not simply harpoon fish and sit in a cave brooding. He quickly discovered that this dark forest contained everything he needed. He built a series of huts out of the native woods. He cultivated various fruit trees. He taught himself to hunt the goats. He domesticated dozens of feral cats—they protected him against the rats and provided him much needed companionship. He took apart his useless rifle and fashioned tools out of it. Recalling what he learned from his father, who had been a shoemaker, he made his own clothes out of animal hides. It was as if he had suddenly come to life and his depression disappeared. He was finally rescued from the island, but the experience completely altered his way of thinking. Years later he would recall his time there as the happiest in his life.

Most of us are like Selkirk when he first found himself stranded—we look at our material resources and wish we had more. But a different possibility exists for us as well—the realization that more resources are not necessarily coming from the outside and that we must use what we already have to better effect. What we have in hand could be research material for a particular book, or people who work within our organization. If we look for more—information, outside people to help us—it won't necessarily lead to anything better; in fact the waiting and the dependence makes us less creative. When we go to work with what is there, we find new ways to employ this material. We solve problems, develop skills we can use again and again, and build up our confidence. If we become wealthy and dependent on money and technology, our minds atrophy and that wealth will not last.

TURN ALL OBSTACLES INTO OPENINGS

The great boxer Joe Louis encountered a tremendous obstacle in the racism of the 1930s. Jack Johnson had preceded Louis as the most famous black boxer of his time. Johnson was supremely skilled and he beat his white opponents with ease, but he was an emotional fighter—encountering hostile crowds that chanted "Kill the nigger" only made him more heated and angry. He found himself in constant trouble and quickly burned out from all the hatred.

Louis was equally talented, but as he perceived it, he could not gloat or show emotion in the ring—that would incite the white audiences and feed into the stereotype of the out-of-control black boxer. And yet a fighter thrives off his emotions, his fighting spirit, and uses this to overwhelm his opponent. Instead of rebelling against this state of affairs or giving up, Louis decided to use it to his advantage. He would show no emotions in the ring. After knocking someone out, he would calmly return to his corner. Opponents and the audience would try to bait him into an emotional response, but he resisted. All of his spirit and anger went into forging this cold and intimidating mask. The racists could not rail against this. He became known as the "Embalmer," and it was enough to see his grim expression when you entered the ring to feel your legs getting weak. In essence, Louis turned this obstacle into his greatest strength.

An opportunist in life sees all hindrances as instruments for power. The reason is simple: negative energy that comes at you in some form is energy that can be turned around—to defeat an opponent and lift you up. When there is no such energy, there is nothing to react or push against; it is harder to motivate yourself. Enemies that hit you have opened themselves up to a counterattack in which you control the timing and the dynamic. If bad publicity comes your way, think of it as a form of negative attention that you can easily reframe for your purposes. You can seem contrite or rebellious, whatever will stir up your base. If you ignore it, you look guilty. If you fight it, you seem defensive. If you go with it and channel it in your direction, you have turned it into an opportunity for positive attention. In general, obstacles force your mind to focus and find ways around them. They heighten your mental powers and should be welcomed.

LOOK FOR TURNING POINTS

Opportunities exist in any field of tension—heated competition, anxiety, chaotic situations. Something important is going on and if you are able to determine the underlying cause, you can create for yourself a powerful opportunity.

Look for any sudden successes or failures in the business world that people find hard to explain. These are often indications of shifts going on under the surface; perhaps someone has inadvertently hit upon a new model for doing things and you must analyze this. Examine the greatest anxieties of those on the inside of any business or industry. Deep changes going on usually register as fear to those who do not know how to deal with them. You can be the first to exploit such changes for positive purposes.

Keep your eye out for any kind of shifts in tastes or values. People in the media or the establishment will often rail against these changes, seeing them as signs of moral decline and chaos. People fear the new. You can turn this into an opportunity by being the first to give some meaning to this apparent disorder, establishing it as a positive value. You are not looking for fads, but deep-rooted changes in people's tastes. One opportunity you can always bank on is that a younger generation will react against the sacred cows of the older generation. If the older set valued spontaneity and pleasure, you can be sure that the younger set will crave order and orthodoxy. By attacking the values of the older generation before anyone else, you can gain powerful attention.

MOVE BEFORE YOU ARE READY

Most people wait too long to go into action, generally out of fear. They want more money or better circumstances. You must go the opposite direction and move before you think you are ready. It is as if you are making it a little more difficult for yourself, deliberately creating obstacles in your path. But it is a law of power that your energy will always rise to the appropriate level. When you feel that you must work harder to get to your goal because you are not quite prepared, you are more alert and inventive. This venture *has* to succeed and so it will.

This has been the way of powerful people from ancient to modern times. When Julius Caesar was faced with the greatest decision of his life—whether to move against Pompey and initiate a civil war or wait for a better moment—he stood at the Rubicon River that separated Gaul from Italy, with only the smallest of forces. Although it seemed insanity to his lieutenants, he judged the moment right. He would compensate for the smallness of his troops with their heightened morale and his own strategic wits. He crossed the Rubicon, surprised the enemy, and never looked back.

When Barack Obama was contemplating a run for president in 2006, almost everyone advised him to wait his turn. He was too young, too much of an unknown. Hillary Clinton loomed over the scene. He threw away all their conventional wisdom and entered the race. Because everything and everyone was against him, he had to compensate with energy, superior strategy, and organization. He rose to the occasion with a masterful campaign that turned all of its negatives into virtues—his inexperience represented change, etc.

Remember: as Napoleon said, the moral is to the physical as three to one—meaning the motivation and energy levels you or your army bring to the encounter have three times as much weight as your physical resources. With energy and high morale, a human can overcome almost any obstacle and create opportunity out of nothing.

Reversal of Perspective

In modern usage, "opportunist" is generally a derogatory term that refers to people who will do anything for themselves. They have no core values beyond promoting their own needs. They contribute nothing to society. This, however, is a misreading of the phenomenon and stems from an age-old elitism that wants to see opportunities kept as privileges for a powerful few. Those from the bottom who dare to promote themselves in any way are seen as Machiavellian, while those already on the top who practice the same strategies are merely smart and resourceful. Such judgments are a reflection of fear.

Opportunism is in fact a great art that was studied and practiced by many ancient cultures. The greatest ancient Greek hero of them all, Odysseus, was a supreme opportunist. In every dangerous moment in his life, he exploited some weakness his enemies left open to trick them and turn the tables. The Greeks venerated him as one who had mastered life's shifting circumstances. In their value system, rigid, ideological people who cannot adapt, who miss all opportunities, are the ones who deserve our scorn—they inhibit progress.

Opportunism comes with a belief system that is eminently positive and powerful—one known to the Stoic philosophers of ancient Rome as *amor fati*, or love of fate. In this philosophy every event is seen as fated to occur. When you complain and rail against circumstances, you fall out of balance with the natural state of things; you wish things were different. What you must do instead is accept the fact that all events occur for a reason, and that it is within your capacity to see this reason as positive. Marcus Aurelius compared this to a fire that consumes everything in its path—all circumstances become consumed in your mental heat and converted into opportunities. A man or woman who believes this cannot be hurt by anything or anyone.

WITHOUT DOUBT, PRINCES BECOME GREAT WHEN THEY OVERCOME DIFFICULTIES AND HURDLES PUT IN THEIR PATH. WHEN FORTUNE WANTS TO ADVANCE A NEW PRINCE . . . SHE CREATES ENEMIES FOR HIM, MAKING THEM LAUNCH CAMPAIGNS AGAINST HIM SO THAT HE IS COMPELLED TO OVERCOME THEM AND CLIMB HIGHER ON THE LADDER THAT THEY HAVE BROUGHT HIM. THEREFORE, MANY JUDGE THAT A WISE PRINCE MUST SKILLFULLY FAN SOME ENMITY WHENEVER THE OPPORTUNITY ARISES, SO THAT IN CRUSHING IT HE WILL INCREASE HIS STANDING.

—Niccolò Machiavelli

CHAPTER 4

Keep Moving—
Calculated Momentum

IN THE PRESENT THERE IS CONSTANT CHANGE AND SO
MUCH WE CANNOT CONTROL. IF YOU TRY TO MICRO-
MANAGE IT ALL, YOU LOSE EVEN GREATER CONTROL IN
THE LONG RUN. THE ANSWER IS TO LET GO AND MOVE
WITH THE CHAOS THAT PRESENTS ITSELF TO YOU—
FROM WITHIN IT, YOU WILL FIND ENDLESS OPPORTUNI-
TIES THAT ELUDE MOST PEOPLE. DON'T GIVE OTHERS
THE CHANCE TO PIN YOU DOWN; KEEP MOVING AND
CHANGING YOUR APPEARANCES TO FIT THE ENVIRON-
MENT. IF YOU ENCOUNTER WALLS OR BOUNDARIES,
SLIP AROUND THEM. DO NOT LET ANYTHING DISRUPT
YOUR FLOW.

||||||||||||||||||||||||||||||||||||||||||||||||||||||||||||||||||||||||||||||||||||||||||||||||||||||||||||||

The Hustler's Flow

||||||||||||||||||||||||||||||||||||||||||||||||||||||||||||||||||||||||||||||||||||||||||||||||||||||||||||||

THE OLD MUSICIANS STAY WHERE THEY ARE AND BE-
COME LIKE MUSEUM PIECES UNDER GLASS, SAFE, EASY
TO UNDERSTAND, PLAYING THAT TIRED OLD SHIT OVER
AND OVER AGAIN. . . . BEBOP WAS ABOUT CHANGE,
ABOUT EVOLUTION. IT WASN'T ABOUT STANDING STILL
AND BECOMING SAFE. IF ANYBODY WANTS TO KEEP
CREATING THEY HAVE TO BE ABOUT CHANGE.

—Miles Davis

When Curtis Jackson first started hustling in the late 1980s, it
was a chaotic world that he entered. Crack cocaine had hit the
streets and turned everything upside down. Now the corner
hustler was unleashed. Moving to wherever there was money
to be made, this new breed of drug dealer had to contend with
hundreds of scheming rivals, the erratic drug addicts, the old-
style gang leaders who were trying to muscle their way back

into the business, and the police who swarmed over the area. It was like the Wild West out there—every man for himself, making up his own rules as he went along.

Some couldn't stand this. They wanted structure, somebody to tell them when to get up and get to work. They didn't last too long in this new order. Others thrived on all the anarchy and freedom. Curtis was of the latter variety.

Then one day everything changed. An old-style gangster nicknamed "the Godfather" made a play for control over the drug traffic of Southside Queens and he succeeded. He installed his son Jermaine in Curtis's neighborhood and the son quickly laid down the law—the family was there to bring order to the business. Jermaine would be selling these purple-top capsules for a cheap price. It would be one size fits all—his capsules or nothing. Nobody could compete with his prices, and any hustler that tried to defy him would be intimidated into submission. They were now all working for Jermaine.

Curtis found this hard to accept. He did not like any kind of authority. He kept trying to get around Jermaine's tight grip on the area by selling his own stuff on the sly, but Jermaine and his team of enforcers kept catching him. Finally they inflicted a good beating on him and he decided it would be wise to surrender—for the time being.

Jermaine liked Curtis's independent spirit and decided to take the youth under his wing, schooling him on what he was up to. He had done some time in prison and had studied business and economics there. He was going to run the crack-cocaine business according to a model inspired by some of the more successful corporations in America. He aimed for control of the local drug business through cheap prices and a complete monopoly on traffic—that was the evolution of all successful enterprises, even the new ones such as Microsoft. He personally hated all the disorder on the streets—it was bad for business and made him uneasy.

One day he drove by in his red Ferrari and invited Curtis to come along for a ride. He drove to the nearby Baisley Projects, which were controlled back then by the Pharaohs, a gang that was heavily invested in the crack trade and notorious for its violent ways. Curtis watched, with growing discomfort, as Jermaine explained to its leaders his plan for the neighborhood. He couldn't have freelancers and gangs operating on the margins of his empire; the Pharaohs would have to fall in line as well, but he'd find a way to make it profitable for them.

The man's arrogance was increasing by the day. Perhaps he would follow this visit up with some violent act to show the Pharaohs he meant what he said, but Curtis had a real bad feeling from that afternoon. Over the next few days, he did whatever he could to avoid running into Jermaine. Sure enough, a week later Jermaine was shot in the head and killed in one of the back alleyways of the hood. Everyone knew who did it and why.

In the months to come, Curtis thought long and hard about what had happened. A part of him had identified with Jermaine. He too had great ambitions and wanted to forge some kind of empire within the hood. With all the competition on the streets, this could never be an easy task. It was natural then for someone like Jermaine to decide that the only way to create this empire was through force and the buildup of a monopoly. But such an effort was futile. Even if he had lasted longer, there were too many people operating on the fringes who resented his takeover and would have done whatever they could to sabotage him. The fiends would have grown tired of his one-size-fits-all approach; they liked variety, even if it was only in the color of the capsules. The police would have taken notice of his large operation and tried to break it up. Jermaine had been living in the past, in ideas cooked up in prison in the 1970s, the grand era of the drug lord. Time had passed him by, and in the ruthless dynamic of the hood, he paid for this with his life.

What was needed was a new skill set, a different mentality for handling the chaos. And Curtis would be the hustler to develop these skills to the maximum. For this purpose, he let go of any desire to dominate an area with one large operation. Instead he started experimenting with four or five hustles at the same time; inevitably one of the angles would work and pay for all the others. He made sure he always had options, room to move in case the police pushed in and cut off one of his access routes. He interacted with the fiends, looking for any changes in their tastes and ways he could appeal to them with some new marketing scheme. He let those who worked for him do things on their own time, as long as they produced results— he wanted as little friction as possible. He never stayed tied to one venture, one partner, or one way of doing things for very long. He kept moving.

The chaos of the streets was part of his flow, something he learned to exploit by working from within it. Operating this way, he slowly accumulated the kind of hustling empire that could surpass what even Jermaine had attempted.

In 2003, Curtis (now known as 50 Cent) found himself thrust into corporate America, working within Interscope Records and dealing with the growing number of businesses that wanted to ally themselves with him. Coming from the streets, with no formal business background, it was natural for him to feel intimidated in this new environment. But within a few months he saw things differently—the new skills he had developed in the hood were more than adequate.

What he noticed about the business executives he dealt with was rather shocking: they operated by these conventions that seemed to have little to do with the incredible changes going on in the business environment. The record industry, for instance, was being destroyed by digital piracy, but the executives could only think of somehow maintaining their monopoly on ownership and distribution; they were incapable of adapting to the changes. They interacted only with themselves—not with their customer base —so their ideas never evolved. They were living in the past, when all of the business models were simple, and control was easy to come by. They had the Jermaine mentality through and through, and in Fifty's mind they would some day suffer a similar fate.

Fifty would stay true to his street strategies—he would opt for fluid positions and room to move. This meant branching out into ventures that were not at all traditional for a rapper—Vitamin Water, a line of books, an alliance with General Motors and Pontiac. These associations seemed disorderly and random, but it was all tied to his compelling image that he continued to shape. He worked five different angles at the same time; if one venture failed, he learned and moved on. The business world was like a laboratory that he would use for constant experimentation and discovery. He would mix and mingle with his employees, up and down the line, and with his audience, allowing them to alter his ideas. The centerpiece of this flow strategy would be the Internet, a chaotic space with endless opportunity for a hustler like himself.

Without knowing exactly where it would lead, he began putting together his own website. At first it was a place to showcase new videos and get feedback from the public. Soon it began to morph into a social network, bringing together his fans from all over the world. This gave him endless space to market his brand and track the changing moods of his audience. His website would continue to evolve like a living organism—he placed no limits on what it could become.

Years later, having moved beyond music into as many varied realms as possible, Fifty could look back on all the people he had left far behind—the record executives, fellow rappers, and business leaders who had gone astray amid all the rapid fluctuations in the early part of the century, a whole gallery of Jermaine types who had no flow. No matter the changes to come, he would continue to thrive in this new Wild West environment, just as he had on the streets.

The Fearless Approach

50 CENT IS A PERSON I CREATED. SOON IT WILL BE
TIME TO DESTROY HIM AND BECOME SOMEBODY ELSE.

—50 Cent

As infants we were surrounded by many things that were unfamiliar and unpredictable—people acting in ways that did
not make sense, events that were hard to figure out. This was
the source of great anxiety. We wanted the world around us
to be more familiar. What was not so predictable became associated in our minds with darkness and chaos, something to
dread. Out of this fear, a desire was born deep inside of us
to somehow gain greater control over the people and events
that eluded our grasp. The only way we knew how to do this
was to grab and hold, to push and pull, exerting our will in
as direct a manner as possible to get people to do what we
wanted. Over the years, this can become a lifelong pattern of
behavior—more subtle as an adult, but infantile at heart.

Every individual we come across in life is unique, with his
or her own energy, desires, and history. But wanting more control over people, our first impulse is generally to try to push
them into conforming to our moods and ideas, into acting in
ways that are familiar and comfortable to us. Every circumstance in life is different, but this elicits that old fear of chaos

and the unknown. We cannot physically make events more predictable, but we can internally create a feeling of greater control by holding on to certain ideas and beliefs that give us a sense of consistency and order.

This hunger for control, common to all of us, is the root of so many problems in life. Staying true to the same ideas and ways of doing things makes it that much harder for us to adapt to the inevitable changes in life. If we try to dominate a situation with some kind of aggressive action, this becomes our only option. We cannot give in, or adapt, or bide our time—that would mean letting go of our grip, and we fear that. Having such narrow options makes it hard to solve problems. Forcing people to do what we want makes them resentful—inevitably they sabotage us or assert themselves against our will. What we find is that our desire to micromanage the world around us comes with a paradoxical effect—the harder we try to control things in our immediate environment, the more likely we are to lose control in the long run.

Most people tend to think of these forms of direct control as power itself—something that shows strength, consistency, or character. But in fact the opposite is the case. They are forms of power that are infantile and weak, stemming from that deep-rooted fear of change and chaos. Before it is too late you need to convert to a more sophisticated, fearless concept of power—one that emphasizes fluidity.

Life has a particular pace and rhythm, an endless stream of changes that can move slowly or quickly. When you try to stop this flow mentally or physically by holding on to things or people, you fall behind. Your actions become awkward because they are not in relation to present circumstances. It is like moving against a current as opposed to using it to propel you forward.

The first and most important step is to let go of this need to control in such a direct manner. This means that you no longer see change and chaotic moments in life as something to fear, but rather as a source of excitement and opportunity. In a social situation in which you want the ability to influence people, your first move is to bend to their different energies. You see what they bring and you adapt to this, then find a way to divert their energy in your direction. You let go of the past way of doing things and adapt your strategies to the ever-flowing present.

Often what seems like chaos to us is merely a series of events that are new and hard to figure out. You cannot make sense of this apparent disorder if you are reactive and fearful, trying to make everything conform to patterns that exist only in your mind. By absorbing more of these chaotic moments with an open spirit, you can glimpse a pattern, a reason why they are occurring, and how you can exploit them.

As part of this new concept, you are replacing the old stalwart symbols of power—the rock, the oak tree, etc.—with that of water, the element that has the greatest potential force in all of nature. Water can adapt to whatever comes its way, moving around or over any obstacle. It wears away rock over time. This form of power does not mean you simply give in to what life brings you and drift. It means that you channel the flow of events in your direction, letting this add to the force of your actions and giving you powerful momentum.

In places like the hood, the concept of flow is more developed than elsewhere. In such an environment, obstacles are everywhere. Those who live there cannot move and make a good living beyond the confines of the hood. If they try to control too many things and become aggressive, they tend to make their lives harder and shorter. The violence they initiate only comes back at them with equal force.

With so many physical limitations, hustlers have learned to develop mental freedom. They cannot let their minds be bothered by all these hindrances. Their thoughts have to keep moving—creating new ventures, new hustles, new directions in music and clothes. That is why trends change so quickly in the hood, which often serves as the engine for new styles in the culture at large. With people, hustlers have to adapt to all of their differences, wearing the mask that is appropriate for each situation, deflecting people's suspicions. (Hustlers are consummate chameleons.) If they can maintain this mental and social fluidity, they can feel a degree of freedom beyond all the physical confinements of the hood.

You too face a world full of obstacles and limitations—a new environment where the competition is more global, complicated, and intense than ever before. Like the hustler, you must find your freedom through the fluidity of your thoughts and your constant inventiveness. This means having a greater willingness to experiment, trying several ventures without fear of failing here or there. It also means constantly looking to develop new styles, new directions you can take, freeing yourself up from any inertia that comes with age. In a world full of people who are too conventional in their thinking, who respect the past far too much, such flow will inevitably translate into power and more room to move.

The fearless types in history all reveal a greater capacity to handle chaos and to use it for their purposes. No greater example of this can be found than Mao Zedong. China in the 1920s was a country on the verge of radical change. The old imperial order that had suffocated China for centuries had finally fallen apart. But fearing the disorder that could be unleashed in such a vast country, the two parties vying for control—the Nationalists and the Communists—opted to try to contain the situation as best they could.

The Nationalists offered the old-style imperial order with a new face. The Communists decided to impose on China the Lenin model—waging a proletariat revolution, centered in urban areas, controlling key cities in the country and enforcing strict adherence to party dogma among its followers. This had worked well in the Soviet Union, creating order in a short period of time, but it had no relevance to China; by the end of the decade this strategy was failing miserably. On the verge of annihilation, the Communists turned to Mao, who had a totally different concept of what to do.

Mao had been raised in a small village, among the country's vast peasant population. As part of his upbringing, he was immersed in the ancient belief systems of Taoism, which saw change as the essence of nature, and conforming to these

changes as the source of all power. In the end, according to
Taoism, you are stronger by having a softness that allows you
to bend and adapt. Mao was not afraid of the vast size and
population of China. The chaos this could represent would
simply become part of his strategy. His idea was to enlist the
help of the peasantry, so that Communist soldiers could blend
in to the countryside like fish in water.

He would not attack city centers or try to occupy any
single position in the country. Instead he would move the
army around, like a vaporous force that would attack and then
disappear, the enemy never knowing where it was coming
from or what it was up to. This guerrilla force would stay in
constant motion, allowing the enemy no breathing space and
giving *them* a sense of chaos.

The Nationalists epitomized the opposite school of fighting, conventional to the core. When Mao finally unleashed on
them his new brand of warfare, they could not adapt. They
held on to key positions, while the Communists encircled
them in the vast spaces of China. The Nationalists' control
narrowed to the point of a few cities, and soon they crumbled
completely in one of the most remarkably swift turnarounds
in military history.

Understand: it is not only what you do that must have flow,
but also how you do things. It is your strategies, your methods of attacking problems, that must constantly be adapted
to circumstances. Strategy is the essence of human action—
the bridge between an idea and its realization in the world.
Too often these strategies become frozen into conventions,

as people mindlessly imitate what worked before. By keeping your strategies attuned to the moment, you can be an agent of change, the one who breaks up these dead ways of acting, gaining tremendous power in the process. Most people in life are rigid and predictable; that makes them easy targets. Your fluid, unpredictable strategies will drive them insane. They cannot foresee your next move or figure you out. That is often enough to make them give way or fall apart.

Keys to Fearlessness

THUS ONE'S VICTORIES IN BATTLE CANNOT BE RE-
PEATED—THEY TAKE THEIR FORM IN RESPONSE TO IN-
EXHAUSTIBLY CHANGING CIRCUMSTANCES. . . . IT CAN
BE LIKENED TO WATER, AS WATER VARIES ITS FLOW AC-
CORDING TO THE FALL OF THE LAND.

—Sun Tzu

All of us have experienced at some point in our lives a feeling of momentum. Perhaps we do something that strikes a chord and we get recognized for it. This positive attention fills us with unusual confidence, which in turn attracts people to us. Now brimming with self-belief, we are able to pull off another good action. Even if it is not so perfect, people will now tend to overlook the rough patches. We have the aura of success about us. So many times in life, one good thing seems to follow another.

This will go on until inevitably we disrupt the flow. Perhaps we overreach with an action that breaks the spell, or we keep repeating the same things and people grow tired of us and move on to someone else. Just as quickly the opposite momentum can afflict us. Our own insecurities start to get in the way; the little imperfections that people overlooked before now seem glaring. We enter a run of bad fortune and feelings of depression render us more and more immobile.

On either end of the spectrum we recognize the phenomenon but we treat it as if it were something mystical, beyond conscious control and explanation. But it is not as mysterious as we might think. In the midst of any run of momentum, we generally feel more open; we allow ourselves to be carried along. The confidence we have when things are going well makes people get out of our way or join our side, giving our actions added force. Sometimes a feeling of urgency—we have to get something done—pushes us to act in a particularly energetic manner, and this starts a good run. This is often accompanied with a feeling that we have little to lose by trying something bold. Perhaps feeling somewhat desperate, we loosen up and experiment.

What ties this all together is that something inside of us opens up and we allow a greater range of motion. Our style becomes freer and bolder, and we move with the current. On the other hand, when a run of momentum ends, it is usually from something we do, a kind of unconscious self-sabotage. We react against this loosening up, out of some fear of where it could lead us. We become conservative and the flow of energy stops, slowly reversing itself into stasis and depression. In many ways, we are the ones in control of this phenomenon, but it does not operate on a conscious enough level.

Understand: momentum in life comes from increased fluidity, a willingness to try more, to move in a less constricted fashion. On many levels it remains something hard to put into words, but by understanding the process, becoming more conscious of the elements involved, you can place your mind in a readied position, better able to exploit any positive movement in your life. Call this calculated momentum. For this purpose you must practice and master the following four types of flow.

MENTAL FLOW

In the time of Leonardo da Vinci's youth (mid-fifteenth century), knowledge had hardened into rigid compartments. In one slot, there was philosophy and scholasticism; in another, the arts, which were considered more like simple crafts; in yet another, science, which was not yet very empirical. On the margins stood all forms of dark knowledge—the arts of the occult.

Da Vinci was the illegitimate son of a notary, and because of this murky social position, he was denied the usual formal education, all of which became a great blessing in disguise. His mind was freed from all the prejudices and rigid categories of thinking that prevailed at the time. He went to serve an apprenticeship in the studio of the great artist Verrocchio. And once he began to learn there the craft of drawing and painting, a process was set in motion that led to the forming of one of the most original minds in the history of mankind.

Knowledge in one field simply opened up in da Vinci an insatiable hunger to learn something else in a related field. The study of painting led to that of design in general, which led to an interest in architecture—from there he flowed to studying engineering; making war machines and strategy; observing animals and the mechanics of motion that could be applied to technology; studying birds and aerodynamics, the anatomy of animals and humans, the relationship between emotions and physiology; and on and on. This incredible stream of ideas even overflowed into areas of the occult. His mind would recognize no boundaries; he sought the connections between all natural phenomena. In this sense, he was ahead of his time and the first real Renaissance man. His discoveries in various fields had a momentum—the intensity of one leading to another. Many could not understand him and thought he was eccentric, even erratic. But great patrons such as King François I of France, and even Cesare Borgia, recognized his genius and sought to exploit it.

Today we have regressed to a point that resembles the pre-Renaissance. Knowledge has once again hardened into rigid categories, with intellectuals shut off in various ghettos. Intelligent people are considered serious by virtue of how deeply they immerse themselves in one field of study, their viewpoint becoming more and more myopic. Someone who crosses these rigid demarcations is inevitably considered a dilettante. After college we are all encouraged to specialize, to learn one thing well and stick to it. We end up strangling ourselves in the narrowness of our interests. With all of these restrictions, knowledge has no flow to it. Life does not have these categories; they are mere conventions that we mindlessly abide by.

Da Vinci remains the icon and the inspiration for a new form of knowledge. In this form, what matters are the connections between things, not what separates them. The mind has a particular momentum itself; when it heats up and discovers something new, it tends to find other items to study and illuminate. All of the greatest innovations in history come from an openness to discovery, one idea leading to another, sometimes coming from unrelated fields. You must develop this spirit and the same insatiable hunger for knowledge. This comes from widening your fields of study and observation, letting yourself be carried along by what you discover. You will find that you will come up with unexpected ideas, the kind that will lead to new practices or novel opportunities. If things run dry in your particular line of work, you have developed your mind along other lines that you can now exploit. Having such mental flow will allow you to constantly think around any obstacle and maintain your career momentum.

EMOTIONAL FLOW

By nature we are emotional creatures. It is how we primarily react to events; only afterwards are we able to see that such emotional responses can be destructive and need to be reined in. You cannot repress this part of human nature, nor should you ever try. It is like a flood that will overwhelm you all the more for your attempts to dam it up. What you want is for these endless emotions that assail you during the day to wash over you, to never hold on to one single emotion for very long. You are able to let go of any kind of obsessive feeling. If someone says something that bothers you, you find a way to move quickly past the feeling—either to excuse what they said, to make it less important, or to forget.

Forgetting is a skill that you must develop in order to have emotional flow. If you cannot help but feel anger or disgust in the moment, make it a point to not let it remain the following day. When you hold on to emotions like that, it is as if you put blinders on your eyes. For that amount of time, you see and feel only what this emotion dictates, falling behind events. Your mind stops on feelings of failure, disappointment, and mistrust, giving you that awkwardness of someone out of tune with the moment. Without realizing it, all of your strategies become infected by these feelings, pushing you off course.

To combat this, you must learn the art of counterbalance. When you are fearful, force yourself to act in a bolder fashion than usual. When you feel inordinate hate, find some object of love or admiration that you can focus on with intensity. One strong emotion tends to cancel out the other and help you move past it.

It might seem that intense feelings of love, hate, or anger can be used to impel you forward on some project, but that is an illusion. Such emotions give you a burst of energy that falls quickly and leaves you as low as you were high. Rather, you want a more balanced emotional life, with fewer highs and lows. This not only helps you keep moving and overcoming petty obstacles, but it also affects people's perceptions of you. They come to see you as someone who has grace under pressure, a steady hand, and they will turn to you as a leader. Maintaining such steadiness will keep that positive flow in motion.

SOCIAL FLOW

Working with people on any level can be a disorderly affair. They bring their differences and own energy to the project, as well as their own agendas. The natural tendency for a leader is to try to tamp down these differences and get everyone on the same page. This seems like the strong thing to do, but in fact it stems from that infantile fear of the unpredictable. And in the end it becomes counterproductive, as those who work for you bring less and less energy to the task. After an initial burst of enthusiasm in your venture, the discontent of those working for you can quickly stifle any momentum you had developed.

Early in his career, the great Swedish director Ingmar Bergman used this more tyrannical approach in dealing with his actors, but he began to be dissatisfied with its results and so decided to experiment with something different. He would sketch out the script for a film, leaving the dialog mostly open. He would then invite his actors to bring their own energy and experiences into the mix, shaping the dialog to fit their emotional responses. This would make the screenplay come alive from within, and sometimes it would require rewriting parts of the plot. In working with the actors on this level, Bergman would enter their spirit, mirroring their energy as a way to get them to relax and open up. He allowed for this more and more as his career evolved, and the results were astonishing.

The actors came to love this, feeling more involved and engaged; they wanted to work with him, and their enthusiasm carried over into their performances, each one better than the last. His films had the feel of something much more lifelike and engaging than those structured around some rigid script. His work became increasingly popular as he went further with this collaborative process.

This should be your model in any venture that involves groups of people. You provide the framework, based on your knowledge and expertise, but you allow room for this project to be shaped by those involved in it. They are motivated and creative, helping to give the project more flow and force. You are not going too far in this process; you set the overall direction and tone. You are simply letting go of that fearful need to make people do exactly as you desire. In the long run, you will find that your ability to gently divert people's energy in your direction gives you a wider range of control over the shape and result of the project.

CULTURAL FLOW

In the 1940s, the great saxophone player Charlie Parker single-handedly revolutionized the world of jazz with his invention of the style known as bebop. But he watched it soon become the convention in jazz, and within a few years he was no longer the revolutionary figure worshipped by hipsters. Younger artists emerged who took his inventions to other levels. This was immensely disturbing to him and he spiraled downward, dying at an early age.

The trumpeter Miles Davis had been a part of Parker's ensemble and he personally witnessed this decline. Davis understood the situation at its core—jazz was an incredibly fluid form of music that underwent tremendous changes in style in short periods of time. Because America did not honor or take care of its black musicians, the ones who found themselves surpassed by a new trend had to suffer a terrible fate, like Parker. Davis vowed to overcome this dynamic. His solution was to never settle on one style. Every four years or so, he would radically reinvent his sound. His audiences would have to catch up with the changes, and most often they did.

It soon became a self-fulfilling prophecy, as he was seen as someone who had his finger on the latest trend, and his new sound would be studied and emulated. As part of this strategy, he would always hire the youngest generation of performers to work with him, harnessing the creativity that comes with youth. In this way, he developed a kind of steady momentum that carried him past the usual decline in a jazz musician's career. He kept this inventiveness up for over thirty years, something unheard of in the genre.

Understand: you exist in a particular cultural moment, with its own flow and style. When you are young you are more sensitive to these fluctuations in taste and so you generally keep up with the present. But as you get older the tendency is for you to become locked in a style that is dead, one that you associate with your youth and its excitement. If enough time passes, your style-lock can become quite ludicrous; you look like a museum piece. Your momentum will grind to a halt as people come to categorize you in a narrow period of time.

Instead you must find a way to periodically reinvent yourself. You are not trying to mimic the latest trend—that will make you look equally ludicrous. You are simply rediscovering that youthful attentiveness to what is happening around you and incorporating what you like into a newer spirit. You are taking pleasure in shaping your personality, wearing a new mask. The only thing you really have to fear is becoming a social and cultural relic.

Reversal of Perspective

In Western culture, we tend to associate strength of character with consistency. People who shift around too much with their ideas and image can be judged as untrustworthy and even demonic. We honor those who are true to the past and certain timeless values. On the other hand, people who challenge and change the prevailing conventions are often viewed as destructive figures, at least while they are alive.

The great Florentine writer Niccolò Machiavelli saw these values of consistency and order as products of a fearful culture and something that should be reversed. In his view, it is precisely our fixed nature, our tendency to hold to one line of action or thought, that is the source of human misery and incompetence. A leader can come to power through acts of boldness, but when the times shift and require something more cautious, he generally will continue with his bold approach. He is not strong enough to adapt; he is a prisoner of his fixed nature. What raised him above others then becomes the source of his downfall.

True figures of power, as Machiavelli saw it, would be people who could shape their own character, call up the qualities that were necessary for the moment, and know how to bend to circumstance. Those who remain true to some idea or value without self-examination often prove to be the worst tyrants in life. They make others conform to dead concepts. They are negative forces, holding back the change that is necessary for any culture to evolve and prosper.

This is how you must operate: you actively work to overcome this fixed nature, deliberately trying a different approach and style than your usual one, to get a sense of a different possibility. You come to view periods of stability and order with mistrust. Something isn't moving in your life and in your mind. On the other hand, moments of change and apparent chaos are what you thrive on—they make your mind and spirit jump to life. If you reach such a point, you have tremendous

power. You have nothing to fear from moments of transition.
You welcome, even create them. Whenever you feel rooted and
established in place, that is when you should be truly afraid.

PEOPLE WISH TO BE SETTLED; ONLY AS FAR AS THEY
ARE UNSETTLED IS THERE ANY HOPE FOR THEM.

—Ralph Waldo Emerson

CHAPTER 5

Know When to Be Bad—Aggression

YOU WILL ALWAYS FIND YOURSELF AMONG THE AG-
GRESSIVE AND THE PASSIVE AGGRESSIVE WHO SEEK
TO HARM YOU IN SOME WAY. YOU MUST GET OVER ANY
GENERAL FEARS YOU HAVE OF CONFRONTING PEOPLE
OR YOU WILL FIND IT EXTREMELY DIFFICULT TO ASSERT
YOURSELF IN THE FACE OF THOSE WHO ARE MORE
CUNNING AND RUTHLESS. BEFORE IT IS TOO LATE YOU
MUST MASTER THE ART OF KNOWING WHEN AND HOW
TO BE BAD—USING DECEPTION, MANIPULATION, AND
OUTRIGHT FORCE AT THE APPROPRIATE MOMENTS. EV-
ERYONE OPERATES WITH A FLEXIBLE MORALITY WHEN
IT COMES TO THEIR SELF-INTEREST—YOU ARE SIMPLY
MAKING THIS MORE CONSCIOUS AND EFFECTIVE.

The Hustler's Setup

[T]HE HUSTLER'S EVERY WAKING HOUR IS LIVED WITH
BOTH THE PRACTICAL AND THE SUBCONSCIOUS
KNOWLEDGE THAT IF HE EVER RELAXES, IF HE EVER
SLOWS DOWN, THE OTHER HUNGRY, RESTLESS FOXES,
FERRETS, WOLVES, AND VULTURES OUT THERE WITH
HIM WON'T HESITATE TO MAKE HIM THEIR PREY.

— Malcolm X

In the summer of 1994, Curtis Jackson returned to Southside
Queens after having served some time in a rehabilitation program for drug offenders. And to his surprise, in the year he
had been away the hustling game had dramatically changed.
The streets were now more crowded than ever with dealers trying to make some money in the crack-cocaine trade.
Having grown weary over the heated rivalries and violence
of the past eight years, the hustlers had settled into a system
where each would have his own corner or two; the drug fiends

would come to them for quick transactions. It was easy and predictable for everyone. No need to fight or push people out of the way or even move around.

When Curtis spread the word that he was looking to get his old crew together and start back where he had left off before rehab, he was met with suspicion and outright hostility. He could ruin the nice system they had in place with his ambitious hustling schemes. He had the feeling they would kill him before he could do anything, just to preserve this new order.

The future suddenly seemed depressing and grim. He had decided months before to find a way out of the drug-dealing racket, but his plans depended on his ability to make some good money and save it so he could segue into a music career. Fitting into this one-corner system would mean he could never earn enough. A few years would go by and he would find it harder and harder to get out. But if he made a play to grab more turf and make some quick money, he would find few allies and many enemies among his fellow hustlers. It was not in their interest that he be allowed to expand his business.

The more Curtis pondered the situation, the angrier he became. It seemed to him that everywhere he turned, people were trying to get in his way, restrain his ambitions, or tell him what to do. They pretended they were trying to keep order, when in fact it was just about getting power for themselves and holding on to it. In his experience, whenever he wanted something in life, he couldn't afford to be nice and submissive; he had to get active and forceful. It would be natural for him to feel a little skittish, coming fresh out of jail and trying to get his old life back together, but what he really had to be afraid of was being stuck and settling for the corner hustler's life. Now was exactly the time to get aggressive, to be bad, and to disrupt this system that was designed only to keep people like him down.

He thought back to the great hustlers he had known in the neighborhood. One of their most successful strategies was the "setup," a variation on the old con game of bait and switch. You distract people with something dramatic and emotional, and while they are not paying attention to you, you grab what you want. He had seen it executed dozens of times, and as he thought about it, he realized he had the material for the perfect distraction.

While in rehab he had befriended the ringleader of a gang of Brooklyn stickup artists. They were notorious for their efficiency and intimidating presence. For the setup, Curtis would lay low for a few weeks, working a corner like everyone else and appearing to go along with the new system. He would then hire these stickup artists on the sly to rob all of the neighborhood hustlers—including Curtis himself—of their jewelry, money, and drugs. They would make several sweeps of the area over the course of a few weeks. As part of the deal, they would keep the money and jewelry from the robberies; Curtis would get the drugs. Nobody would suspect his involvement.

In the weeks to come he watched with amusement as the sudden appearance of the stickup artists in his neighborhood caused panic among the hustlers, some of whom were his friends. He pretended to share their distress. These Brooklyn gangstas were not to be messed with. Almost overnight, the dealers' whole way of life was disrupted: they were forced now to carry guns for protection, but this created a new set of problems. The police were everywhere, making random checks, and to be caught loitering with a gun would mean solid prison time. The hustlers could no longer simply stand on the street corner and wait for the drug fiends to come to them. They had to keep in constant motion to avoid the police; for some, getting called on their beepers was the only way to arrange a deal. Everything became more complicated and business slowed down.

The old model, tight and static, had been exploded, and now Curtis moved into the breach with some new-colored capsules he packaged and sold to the fiends. Sometimes he included in the sales some free capsules, which happened to be the drugs he had accumulated from the robberies. The fiends began to flock to him, while the other hustlers were too upset to notice the trick that had been played. By the time they had figured it out, it was too late. Curtis had expanded his business and he was well on his way to buying his freedom.

Several years later, Curtis (now known as 50 Cent) had carved out a path towards a career as a rapper. He had a deal with Columbia Records and the future looked reasonably bright. But Fifty was not one to be duped by the usual dreams. He quickly saw that there was only so much room for the top performers who could bank on a solid career in this business. He, along with everyone else, was fighting for crumbs of attention; the artists might get temporary success with a hit here or there, but it wouldn't last, and they had no power to alter the dynamic. What was worse, Fifty had made some enemies in the business—he was an ambitious hustler with talent. There were people who mistrusted and feared him. They worked behind the scenes to make sure he would not get far in the industry.

As Fifty had learned, talent and good intentions are never enough in this world; you need to be fearless and strategic. When you face people's indifference or outright hostility you have to get aggressive and push them out of your way by any means necessary, and not worry about some people disliking you. In this case he looked for any opportunity to make such a bold move, and one evening a chance encounter provided this for him.

At a club in Manhattan, Fifty was talking with a friend from the neighborhood when he saw the rapper Ja Rule staring in his direction. Several weeks before, Fifty's friend had robbed Ja Rule of some jewelry in broad daylight; Fifty expected Ja to come over and cause some trouble. Instead he looked away and decided to ignore them. This was rather shocking. Ja Rule was then one of the hottest rappers in the business; he had built his reputation on being a gangsta from Southside Queens, his lyrics reflecting his tough-guy image. He and his record label, Murder Inc., had allied themselves with Kenneth "Supreme" McGriff, former head of the Supreme Team, a gang that had dominated the New York drug business in the 1980s with its ruthless tactics. Supreme gave them street credibility, and Murder Inc. gave Supreme an entrée into the music business, something legitimate to distance himself from his dark past.

No real hustler or gangsta would ever ignore the man who had robbed him so brazenly. What this meant to Fifty was that Ja was fake, his lyrics and image all a show to make money. He was arrogant yet insecure. As he contemplated this, the idea for a masterful setup took shape in his mind, one that would draw attention and help catapult him past all of those who stood in his way.

In the weeks to come Fifty began releasing diss tracks that took aim at Ja Rule, painting him as a studio gangsta who rapped about things that he had never experienced. Ja must have been annoyed but he did not respond to Fifty's taunts. He was clearly too big to concern himself with small fry. Fifty's next move, however, could not be ignored. He released a song that detailed the activities of the most notorious gang leaders (including Supreme) in Southside Queens in the 1980s. As the song became popular on the streets, it brought Supreme the kind of attention he was trying to avoid, now that he was going legitimate. This made him angry and suspicious—what might Fifty do next? He pressured Ja Rule to go after and destroy this upstart before he went too far.

Ja was now forced into going after him. He tried whatever he could to shut Fifty up: he spread nasty rumors about Fifty's past and attempted to block any record deals he might have; at one point, finding Fifty in the same recording studio he was at, Ja and his cohorts started a brawl. Ja wanted to intimidate Fifty with his muscle and reputation, but this only made Fifty increase the pace of diss tracks that he released. He wanted to push all of Ja's buttons—make him angry and insecure, burning for revenge. He'd stay cool and strategic, while Ja would lose control. To this purpose, he called Ja a "wanksta," a wannabe gangsta. He parodied his singing style, his lyrics, everything about his supposed tough-guy image. The songs were hard driving, biting, and humorous.

Slowly but surely, Ja became more and more furious as these songs made it to the radio and journalists peppered him with questions about Fifty. He had to prove his toughness, that he was no wanksta, so he released his own diss tracks. These songs were not witty, however, only violent and vicious. Without realizing it, he had become defensive and not very entertaining.

Fifty's first record came out at about the same time as one of Ja's, and its sales far eclipsed those of his rival's. Now he was the star and Ja began to fade from the scene. Befitting his new role, Fifty stopped the attacks, almost out of pity for his former rival. Ja had served his purpose and it was better to leave him to oblivion.

The Fearless Approach

THE WAY I LEARNED IT, THE KID IN THE SCHOOL YARD WHO DOESN'T WANT TO FIGHT ALWAYS LEAVES WITH A BLACK EYE. IF YOU INDICATE YOU'LL DO ANYTHING TO AVOID TROUBLE, THAT'S WHEN YOU GET TROUBLE.

—50 Cent

Life involves constant battle and confrontation. This comes on two levels. On one level, we have desires and needs, our own interests that we wish to advance. In a highly competitive world, this means we must assert ourselves and even occasionally push people out of position to get our way. On the other level, there are always people who are more aggressive than we are. At some point they cross our path and try to block or harm us. On both levels, playing offense and defense, we have to manage people's resistance and hostility. This has been the human drama since the beginning of history and no amount of progress will alter this dynamic. The only thing that has changed is how we handle these inevitable moments of friction in our lives.

In the past, people had a greater taste for battle. We can read signs of this in all kinds of social behavior. At the theater, for instance, it was common practice for nineteenth-century audiences in Europe and America to verbally express their disapproval of the actors or the play, yelling, hissing, or throwing things onto the stage. Fights would often break out in the theater over differences of opinion; it was not cause for concern but part of the appeal. In political campaigns, it was accepted as normal that partisans of various parties would confront each other in the streets over their divergent interests. Democracy was considered vibrant by allowing such public disagreements, a kind of safety valve for human aggression.

Now we tend to find the opposite. We are generally much more skittish when it comes to confrontation. We often take it personally if someone overtly disagrees with us or expresses an opinion contrary to our own. We are also more afraid of saying something that could possibly offend those around us, as if their egos were too fragile. The culture tends to elevate as its ideal a spirit of cooperation; being democratic and fair means getting along with others, fitting in, and not ruffling feathers. Conflict and friction are almost evil; we are encouraged to be deferential and agreeable. Nevertheless, the human animal retains its aggressive impulses and all that happens is that many people channel this energy into passive-aggressive behavior, which makes everything more complicated.

In such an atmosphere, we all pay a price. When it comes to the offensive side of power, in which we are required to take forceful and necessary action to advance our interests, we are often hesitant and uncertain. When dealing with the aggressors and passive aggressors around us we can be quite naive; we want to believe that people are basically peaceful and desire the same things as ourselves. We often learn too late that this is not the case. This inability to deal with what is inevitable in life is the cause of so many problems. We work to postpone or avoid conflicts, and when they reach a point where we can no longer play such a passive game, we lack the experience and the habit of meeting them head on.

The first step in overcoming this is to realize that the ability to deal with conflict is a function of inner strength versus fear, and that it has nothing to do with goodness or badness. When you feel weak and afraid, you have the sense that you cannot handle any kind of confrontation. You might fall apart or lose control or get hurt. Better to keep everything smooth and even. Your main goal then is to be liked, which becomes a kind of defensive shield. (So much of what passes for good and nice behavior is really a reflection of deep fears.)

What you want instead is to feel secure and strong from within. You are willing to occasionally displease people and you are comfortable in taking on those who stand against your interests. From such a position of strength, you are able to handle friction in an effective manner, being bad when it is appropriate.

This inner strength, however, does not come naturally. What is required is some experience. This means that in your daily life you must assert yourself more than usual—you take on an aggressor instead of avoiding him; you strategize and push for something you want instead of waiting for someone to give it to you. You will generally notice that your fears have exaggerated the consequences of this kind of behavior. You are sending signals to others that you have limits they cannot cross, that you have interests you are willing to defend or advance. You will find yourself getting rid of this constant anxiety about confronting people. You are no longer tied to this false niceness that wears on your nerves. The next battle will be easier. Your confidence in handling such moments of friction will grow with each encounter.

In the hood, people don't have the luxury of worrying about whether people like them. Resources are limited; everyone is angling for power and trying to get what they can. It is a rough game and there is no room for being naive or waiting for good things to happen. You learn to take what you need and feel no guilt about it. If you have dreams and ambitions, you know that to realize them you have to get active, make some noise, bruise a few people in your path. And you expect others to do the same to you. It is human nature, and instead of complaining you simply must get better at protecting yourself.

We all face a similar competitive dynamic—people all around us are struggling to advance their interests. But because our fights are more subtle and veiled, we tend to lose sight of the harsh aspects of the game. We are often too trusting—in others, in a future that will bring good things. We could use some of the toughness and realism that people who grow up in pressurized environments have. A simple line can be drawn— we all have ambitions and large goals for ourselves. We are either waiting for some perfect moment to realize them, or we are taking action in the present. This action requires some aggressive energy channeled in a smart manner and the willingness to displease a person or two who gets in our way. If we are waiting and settling for what we have, it is not because we are good and nice but because we are fearful. We need to get rid of the fear and guilt we might have for asserting ourselves. It serves no purpose except to keep us down.

The fearless types in history have often had to face a lot of hostility in their lives, and in doing so they invariably discover the critical role that one's attitude plays in thwarting people's aggression. Look at Richard Wright, the first bestselling African American writer in U.S. history. His father abandoned his mother shortly after Wright's birth in 1908, and Wright knew only poverty and starvation as a child. His uncle, with whom they lived, was lynched by a white mob, and his family (Wright and his mother and brother) was forced to flee from Arkansas and wander across the South. When his mother fell ill and became an invalid, he was shunted from family to family, even spending time in an orphanage. The family members who took him in, themselves poor and frustrated, beat him incessantly. His classmates at school, sensing he was different (he liked to read books and was shy), taunted and ostracized him. At work, his white employers subjected him to endless indignities, such as beatings and dismissals from the job for no apparent reason.

These experiences created in him intricate layers of fear. But as he read more books about the wider world and thought more deeply, a different spirit rose inside of him—a need to rebel and not accept the status quo. When an uncle threatened to beat him over a triviality, he decided he had had enough. Although just a child, he clutched two razor blades in his hands and told the uncle he was prepared to go down fighting. He was never bothered by that uncle again. Seeing the power he had with such an attitude, he now made it something more calculated and under control. When conditions at work became impossible, he would leave the job—a sign of impertinence to the white employers, who spread word of this around town. He didn't care if people thought he was different—he was proud of it. Feeling like he was going to be trapped in Jackson, Mississippi, for the rest of his life and yearning to escape to the North, he became a criminal for the first and last time in his life, stealing enough to pay his way out of town. He felt more than justified in doing this.

This spirit permeated his life to the very end. As a successful writer now living in Chicago, he felt that his novels were being misread by the white public—they invariably found a way to soften his message about racial prejudices, to see what they wanted to see in his work. He realized he had been holding back, tailoring his words to appeal to them. He had to rise again above this fear of pleasing others and write a book that could not be misread, that would be as bleak as the life he had known. This became *Native Son*, his most famous and successful novel.

What Wright had discovered was simple: when you submit in spirit to aggressors or to an unjust and impossible situation, you do not buy yourself any real peace. You encourage people to go further, to take more from you, to use you for their own purposes. They sense your lack of self-respect and they feel justified in mistreating you. When you are humble, you reap the wages of humility. You must develop the opposite—a fighting stance that comes from deep within and cannot be shaken. You force some respect.

This is how it is in life for everyone: people will take from you what they can. If they sense that you are the type of person who accepts and submits, they will push and push until they have established an exploitative relationship with you. Some will do this overtly; others are more slippery and passive aggressive. You must demonstrate to them that there are lines that cannot be crossed; they will pay a price for trying to push you around. This comes from your attitude—fearless and always prepared to fight. It radiates outward and can be read in your manner without you having to speak a word. By a paradoxical law of human nature, trying to please people less will make them more likely in the long run to respect and treat you better.

Keys to Fearlessness

[F]OR HOW WE LIVE IS SO FAR REMOVED FROM HOW WE OUGHT TO LIVE, THAT HE WHO ABANDONS WHAT IS DONE FOR WHAT OUGHT TO BE DONE, WILL RATHER LEARN TO BRING ABOUT HIS OWN RUIN THAN HIS PRESERVATION. A MAN WHO WISHES TO MAKE A PROFESSION OF GOODNESS IN EVERYTHING MUST NECESSARILY COME TO GRIEF AMONG SO MANY WHO ARE NOT GOOD. THEREFORE IT IS NECESSARY FOR A PRINCE, WHO WISHES TO MAINTAIN HIMSELF, TO LEARN HOW NOT TO BE GOOD, AND TO USE THIS KNOWLEDGE AND NOT USE IT, ACCORDING TO THE NECESSITY OF THE CASE.

—Niccolò Machiavelli

When it comes to morality, almost all of us experience a split in our consciousness. On one hand, we understand the need to follow certain basic codes of behavior that have been in place for centuries. We try our best to live by them. On the other hand, we also sense that the world has become infinitely more competitive than anything our parents or grandparents have known. To get ahead in this world we must be willing to occasionally bend that moral code, to play with appearances, to hedge the truth and make ourselves look better, to manipulate a person or two to secure our position. The culture at large reflects this division. It emphasizes values of cooperation and decency, while titillating us constantly in the media with endless stories of those who have risen to the top by being bad and ruthless. We are both drawn to and repulsed by these stories.

This split creates an ambivalence and awkwardness in our actions. We are not very good at being either good or bad. When we do the manipulative acts that are necessary, it is with half a heart and some guilt. We are not sure how to operate in this way—when to play the more aggressive role, or how far to go.

The great sixteenth-century Florentine writer Niccolò Machiavelli noticed a similar phenomenon in his day, on a different level. Italy had splintered into several city-states that were constantly intriguing for power. It was a dangerous, complicated environment for a leader. In facing a rival state, a prince had to be extremely careful. He knew that these rivals would do anything to advance their interests, including cutting deals with others to isolate or destroy him. He had to be ready to attempt any kind of maneuver to protect his state. At the same time he was imbued with Christian values. He had to juggle two codes of behavior—one for private life and another for the game of power. This made for awkwardness. Nobody really defined the moral parameters for how to defend and advance his state. If he became too aggressive, he would look bad on the world stage and suffer for it. If he was too good and gentle, his state could be overrun by a rival, bringing misery for his citizenry.

For Machiavelli, the problem wasn't a leader adjusting his morality to the circumstance—everyone does that. It was that he did not do it well. Too often he would be aggressive when he needed to be cunning, or vice versa. He would not recognize in time the once friendly state that was now plotting against him, and his response would be too desperate. When a venture succeeds, people tend to overlook some of the nasty tactics you were forced to use; when a venture fails, those same tactics become scrutinized and condemned.

A prince or leader must first and foremost be effective in his actions and to do so he must master the art of knowing when and how to be bad. This requires some fearlessness and flexibility. When the situation calls for it, he must be the lion—aggressive and direct in protecting his state, or grabbing something to secure its interests. At other times, he has to be the fox—getting his way through crafty maneuvers that disguise his aggression. And often he must play the lamb— the meek, deferential, and good creature exalted in culture. He is bad in the right way, calibrated to the situation, and careful to make his actions look justified to the public, reserving his nastier tactics for behind the scenes. If he masters the art of being bad, he uses it sparingly and he creates more peace and power for his citizens than the awkward prince who tries to be too good.

This should be the model for us as well. We are all now princes competing with thousands of rival "states." We have our aggressive impulses, our desires for power. These impulses are dangerous. If we act upon them unconsciously or awkwardly, we can create endless problems for ourselves. We must learn to recognize the situations that require assertive (yet controlled) actions, and which mode of attack (fox or lion) is suitable.

The following are the most common foes and scenarios that you will encounter in which some form of badness is required to defend or advance yourself.

AGGRESSORS

By 1935, there were some on the left in the United States who had grown discontented with President Franklin Delano Roosevelt's reforms, known as the New Deal. They believed these reforms were not working fast enough. They decided to band together to form what would later be known as the Union Party, to galvanize this discontent. They were going to run against FDR in 1936, and the threat was very real that they would gain enough support to throw the election to the Republicans. Within this group was Huey Long, the great populist senator from Louisiana, and Father Charles Coughlin, the Catholic priest who had a popular radio program. Their ideas were vaguely socialistic and appealed to many who felt disenfranchised during the Great Depression. Their attacks on FDR began to have effect; his poll numbers went down. Feeling emboldened, they became even more aggressive and relentless in their campaign.

In the midst of this, FDR remained mostly silent, letting them fill the air with their charges and threats. His advisers panicked; they felt he was being too passive. But for Roosevelt it was part of a plan—he felt certain that the public would grow tired of their shrill attacks as the months went on; he sensed that the factions within the Union Party would begin to fight among themselves as the election neared. He ordered his surrogates to not attack these men.

At the same time, he went to work behind the scenes. In Louisiana, he fired as many Long supporters working for the government as possible and replaced them with those on his side. He launched detailed investigations into the senator's dubious financial affairs. With Coughlin, he worked to isolate him from other notable Catholic priests, making him look like a fringe radical. He introduced laws that forced Coughlin to get an operating permit to broadcast his shows; the government found reasons to deny his requests and temporarily silenced him. All of this served to confuse and frighten FDR's foes. As he had predicted, the party began to splinter and the public lost interest. Roosevelt won the 1936 election in an unprecedented landslide.

FDR had understood the basic principle in squaring off against aggressors who are direct and relentless. If you meet them head on, you are forced to fight on their terms. Unless you happen to be an aggressive type, you are generally at a disadvantage against those who have simple ideas and fierce energy. It is best to fight them in an indirect manner, concealing your intentions and doing what you can behind the scenes—hidden from the public—to create obstacles and sow confusion. Instead of reacting, you must give aggressors some space to go further with their attacks, getting them to expose themselves in the process and provide you plenty of juicy targets to hit. If you become too active and forceful in response, you look defensive. You are playing the fox to their lion—remaining cool and calculating, doing whatever you can to make them more emotional and baiting them to fall apart through their own reckless energy.

PASSIVE AGGRESSORS

These types are masters at disguise. They present themselves as weak and helpless, or highly moral and righteous, or friendly and ingratiating. This makes them hard to pick out at first glance. They send all kinds of mixed signals—alternating between friendly, cool, and hostile—creating confusion and conflicting emotions. If you try to call them on their behavior, they use this confusion to make you feel guilty, as if you were the one who was the source of the problem. Once you are drawn into their dramas, with your emotions engaged, it can be very difficult to detach yourself. The key is recognizing them in time to take appropriate action.

When the Grand Duchess Catherine (future empress of Russia, Catherine the Great) met her husband-to-be, Peter, she felt he was an innocent child at heart. He continued to play with toy soldiers and had a petulant, moody temperament. Then shortly after their marriage in 1745, she began to detect a different side to his character. In private they got along well enough. But then she would hear from secondhand sources all kinds of nasty stories about how he had regretted their marriage and how he preferred her chambermaid. What was she to believe—these stories or his geniality when they were together? After he became Czar Peter III, he would graciously invite Catherine to visit him in the morning, but then he would ignore her. When the royal gardener stopped delivering her favorite fruits, she found out it was on his orders. Peter was doing everything he could to make her life miserable and humiliate her in subtle ways.

Fortunately Catherine figured out early on that he was a master manipulator. His childish exterior was clearly there to distract attention from his petty, vindictive core. His goal, she believed, was to bait her into doing something rash that would give him an excuse to isolate or get rid of her. She decided to bide her time, be as gracious as possible, and win over some key allies in the court and the military, many of whom had come to despise the czar.

Finally, certain of her allies' support, she instigated a coup that would get rid of him once and for all. When it became clear that the military had sided with Catherine and that he was to be arrested, Peter started to beg and plead with her: he would change his ways, and they would rule together. She did not reply. He sent another message saying he would abdicate, if only he could return peacefully to his own estate with his mistress. She refused to bargain. He was arrested and soon thereafter murdered by one of the coup intriguers, perhaps with the approval of the empress.

Catherine was a classic fearless type. She understood that with passive aggressors you must not get emotional and drawn into their endless intrigues. If you respond indirectly, with a kind of passive aggression yourself, you play into their hands—they are better at this game than you are. Being underhanded and tricky only spurs on their insecurities and intensifies their vindictive nature. The only way to treat these types is to take bold, uncompromising action that either discourages further nonsense or sends them running away. They respond only to power and leverage. Having allies higher up the chain can serve as a means of blocking them. You are playing the lion to their fox, making them afraid of you. They see there will be real consequences if they continue their behavior in any form.

To recognize such types, look for extremes in behavior that are not natural—too kind, too ingratiating, too moral. These are most likely disguises that are worn to deflect attention from their true nature. Better to be proactive and take precautionary measures the moment you feel they are trying to get into your life.

UNJUST SITUATIONS

Some time in the early 1850s, Abraham Lincoln came to the conclusion that the institution of slavery was the great stain on our democracy and had to be eliminated. But as he surveyed the political landscape he became concerned: the politicians on the left were too noisy and righteous—in their fervor to promote abolition, they would polarize the country and the slaveholders could easily exploit these political divisions to maintain their way of life for decades. Lincoln was the consummate realist—if your goal is to end an injustice, you have to aim for results, and that requires being strategic and even deceptive. To end slavery he would be willing to do almost anything.

He decided he was the politician best suited for this cause. His first step was to present a moderate front to the public in the 1860 campaign and after his election to the presidency. He gave the impression that his main goal was to maintain the Union and to gradually phase slavery out of existence through a policy of containment. When war became inevitable in 1861, he decided to lay a clever trap for the South, baiting them into an attack on Fort Sumter that would force him to declare war. This made it seem that the North was the victim of aggression. All of these maneuvers were designed to keep his support in the North relatively unified—to oppose him was to oppose his efforts to defeat the South and maintain the Union, the slavery issue slipping into the background. This unified front on his side made it almost impossible for the enemy to play political games.

As the tide of the war turned in favor of the North, he gradually shifted to more radical positions (stated in the Emancipation Proclamation and his Gettysburg Address), knowing he had more leeway to reveal his real goals and act on them. Leading the North to victory in the war, he had even more room to continue his campaign. In sum, to defeat slavery Lincoln was prepared to publicly manipulate opinion by concealing his intentions, and to practice outright deceit in his political maneuverings. This required great fearlessness and patience on his part, as almost everyone misread his intentions and criticized him as an opportunist. (Some still do.)

In facing an unjust situation, you have two options. You can loudly proclaim your intentions to defeat the people behind it, making yourself look good and noble in the process. But in the end, this tends to polarize the public (you create one hardened enemy for every sympathizer won over to the cause), and it makes your intentions obvious. If the enemy is crafty, this makes it almost impossible to defeat them. Or, if it is results you are after, you must learn instead to play the fox, letting go of your moral purity. You resist the pull to get emotional, and you craft strategic maneuvers designed to win public support. You shift your position to suit the circumstance, baiting the enemy into actions that will win you sympathy. You conceal your intentions. Think of it as war—short of unnecessary violence, you are called to do whatever it takes to defeat the enemy. There is no nobility in losing if an injustice is allowed to prevail.

STATIC SITUATIONS

In any venture, people quickly create rules and conventions that must be followed. This is often necessary to instill some discipline and order. But most often these rules and conventions are arbitrary—they are based on something that was successful in the past but might have little relevance to the present. They are often instruments for those in power to maintain their grip and keep the group unified. If this goes on long enough, they become stultifying and crowd out any new ways for doing things. In such a situation, what is called for is the total destruction of these dead conventions, creating space for something new. In other words you must be the complete lion, as bad as can be.

This was how several important black jazz musicians—such as Charlie Parker, Thelonious Monk, and Dizzy Gillespie—responded to the musical conventions that had hardened in the early 1940s. From its more freewheeling earlier days, jazz had become co-opted by white performers and audiences. The sound that became popular—big band, swing—was more controlled and regimented. To make any money in the business you had to play by the rules and perform these popular genres. But even those black musicians who followed the conventions were still paid considerably less than their white counterparts. The only way around this oppressive situation was to destroy it with a completely new sound, in this case with something that later became known as bebop. This new genre went against all the current conventions. The music was wild and improvisatory. As it became popular, it gave these musicians some space to perform on their own terms and some control over their careers. Now the static situation was broken and the field was left open to the great jazz innovations of the 1950s and '60s.

In general, you must be less respectful of the rules that other people have established. They do not necessarily fit the times or your temperament. And there is great power to be had by being the one to initiate a new order.

IMPOSSIBLE DYNAMICS

Sometimes in life you find yourself in a negative situation that cannot be improved no matter what you do. You might find yourself working for people who are irrational. Their actions seem to serve no purpose apart from imposing their power and making you miserable. Everything you do is wrong. Or it could be a relationship in which you are constantly forced to rescue a person. This usually involves types who present themselves as weak victims in need of attention and assistance. They stir up a lot of drama around them. No matter what you do, the need to be rescued keeps recurring.

You can recognize this dynamic by your emotional need to somehow solve the problem, mixed with your complete frustration in finding any kind of reasonable answer. In truth the only viable solution is to terminate the relationship—no arguing, no bargaining, no compromising. You leave the job (there are always others); you leave the person who is tormenting you with as much finality as possible. Resist the temptation to feel any guilt. You need to create as much distance as possible, so they cannot inveigle these emotions into you. They must become dead to you so you can go on with your life.

Reversal of Perspective

The problem with confrontational moments, and why we often seek to avoid them, is that they churn up a lot of unpleasant emotions. We feel personally aggrieved that someone is trying to hurt or harm us. This makes us wonder about ourselves and feel insecure. Did we deserve this in some way? If we go through a few of these unpleasant moments, we become increasingly skittish. But this is really a problem of perception. In our own inner turmoil we tend to exaggerate the negative intentions of our opponents. In general we take conflicts far too personally. People have problems and traumas that they carry with them from their childhood on. Most often when they do something to harm or block us, it really is not directed at us personally. It comes from some unfinished business from the past, or deep insecurities. We happen to cross their path at the wrong moment.

It is essential that you develop the reverse perspective: life naturally involves conflicting interests; people have their own issues, their own agendas, and they collide with yours. Instead of taking this personally or concerning yourself with people's intentions, you must simply work to protect and advance yourself in this competitive game, this bloody arena. Focus your attention on their maneuvers and how to deflect them. When you have to resort to something that isn't conventionally moral, it is just another maneuver you are executing in the game—nothing to feel guilty about. You accept human nature and the idea that people will resort to aggression. This calm, detached perspective will make it that much easier to design the perfect strategy for blunting their aggression. With your emotions unscathed by these battles, you will grow accustomed to them and will even take some pleasure in fighting them well.

IN THE RING, OUR OPPONENTS CAN GOUGE US WITH THEIR NAILS OR BUTT US WITH THEIR HEADS AND LEAVE A BRUISE, BUT WE DON'T DENOUNCE THEM FOR IT OR GET UPSET WITH THEM OR REGARD THEM FROM THEN ON AS VIOLENT TYPES. WE JUST KEEP AN EYE ON THEM . . . NOT OUT OF HATRED OR SUSPICION. JUST KEEPING A FRIENDLY DISTANCE. WE NEED TO DO THAT IN OTHER AREAS. WE NEED TO EXCUSE WHAT OUR SPARRING PARTNERS DO, AND JUST KEEP OUR DISTANCE—WITHOUT SUSPICION OR HATRED.

—Marcus Aurelius

CHAPTER 6

Lead from the Front— Authority

IN ANY GROUP, THE PERSON ON TOP CONSCIOUSLY OR UNCONSCIOUSLY SETS THE TONE. IF LEADERS ARE FEARFUL, HESITANT TO TAKE ANY RISKS, OR OVERLY CONCERNED FOR THEIR EGO AND REPUTATION, THEN THIS INVARIABLY FILTERS ITS WAY THROUGH THE ENTIRE GROUP AND MAKES EFFECTIVE ACTION IMPOSSIBLE. COMPLAINING AND HARANGUING PEOPLE TO WORK HARDER HAS A COUNTERPRODUCTIVE EFFECT. YOU MUST ADOPT THE OPPOSITE STYLE: IMBUE YOUR TROOPS WITH THE PROPER SPIRIT THROUGH YOUR ACTIONS, NOT WORDS. THEY SEE YOU WORKING HARDER THAN ANYONE, HOLDING YOURSELF TO THE HIGHEST STANDARDS, TAKING RISKS WITH CONFIDENCE, AND MAKING TOUGH DECISIONS. THIS INSPIRES AND BINDS THE GROUP TOGETHER. IN THESE DEMOCRATIC TIMES, YOU MUST PRACTICE WHAT YOU PREACH.

The Hustler King

NO MAN CAN PROPERLY COMMAND AN ARMY FROM
THE REAR. HE MUST BE AT THE FRONT . . . AT THE
VERY HEAD OF THE ARMY. HE MUST BE SEEN THERE,
AND THE EFFECT OF HIS MIND AND PERSONAL ENERGY
MUST BE FELT BY EVERY OFFICER AND MAN PRESENT
WITH IT . . . .

—General William T. Sherman

By the spring of 1991, young Curtis Jackson had proved himself to be one of the savviest hustlers in the neighborhood. His
pool of repeat customers had increased to a point where he had
to hire his own crew to keep up with their demand. But as
he knew, nothing good ever lasts too long in the hood. Just as
Curtis was making plans to expand his business, an older hustler named Wayne began to make threatening gestures towards
him. Wayne had recently returned to the streets from prison; he

was determined to make as much money as fast as possible and then dominate the local drug trade. Curtis, it seemed, was his main rival. He tried to intimidate the younger hustler, warning him that he better curtail his operations or pay a price. Curtis ignored him. Then Wayne decided to up the ante: he sent out word on the street that he was going to have Curtis killed.

Curtis had seen this happen before and he knew what would happen next. Wayne would never do the job himself—he could not risk a return to prison. Instead he was banking on the fact that some young kid would hear of his desire to kill Curtis and, eager to gain some street credibility, would take it upon himself to do the dirty work. Sure enough, a few days after hearing of Wayne's intentions, Curtis noticed a young kid named Nitty trailing him on the streets. He felt certain that Nitty was the one planning to do the hit, and it would happen soon.

This was the depressing dynamic of hustling in the hood: the more success a hustler had, the more he attracted the wrong kind of attention. Unless he inspired some fear and terror, rivals would keep coming at him, trying to take what he had and continually threatening his position on the streets. Once that started to happen, the once successful dealer would find himself drawn into a cycle of violence, reprisals, and time in the pen.

There were a few hustlers, however, who had somehow managed to rise above this dynamic. In the hood, they were like kings—just hearing their names or seeing them on the street would elicit a gut reaction, a mix of fright and admiration. What elevated them above others was a series of actions they had taken in the past that demonstrated they were fearless and smart. Their maneuvers would be unpredictable and all the more terrifying for it. If people thought of challenging them, they would quickly remember what these types had done in other circumstances and back off. All of this would give them an aura of power and mystery. Instead of challengers on all sides, they would have disciples ready to follow them as far as they wanted. If Curtis saw himself as the kingly type, it was time now to show it to others, as dramatically as possible.

With death staring him in the face, he worked to control his emotions and thought long and hard about the dilemma that Wayne had posed. If he came after Wayne to kill him first, Wayne would be ready and would have the perfect excuse to kill Curtis in self-defense. If instead he went after Nitty and killed him, the police would catch Curtis and he would end up in prison for a long time, an equally fortunate result for Wayne. And if he did nothing, Nitty would finish him off. But Wayne's strategy had a fatal flaw—his fear of doing the job with his own hands. He was no king himself, but just another frightened hustler pretending to be tough. Curtis would come after him from an unexpected angle and turn everything around.

Without wasting any more time, he asked a member of his crew named Tony to accompany him that afternoon. Together they surprised Nitty on the street, and while Tony held him, Curtis slashed the kid in the face with a razor blade. He did it just deep enough to send him screaming to the hospital, and to leave a nice scar for a while. Then a few hours later, he and Tony found Wayne's empty car and shot it up—an ambiguous message that meant either they hoped he was inside, or they were taunting him to come out and attack them in the open.

The following day, the dominoes fell just as Curtis thought they would. Nitty sought out Wayne, expecting that the two of them would then go together to exact revenge on Curtis— after all, Wayne had been attacked as well. Wayne, however, still insisted the kid do it alone. Now Nitty could see through the game—he was just the patsy to do the dirty work, and Wayne was not as tough as he had made himself out to be. Nitty would have nothing more to do with him, but he was also too afraid to take on Curtis by himself. He decided he could live with the scar. Wayne was now in a delicate position. If he asked someone else to do the job it would start to look like what it was—a man too scared to do it himself. Better to let the whole thing just go away.

In the days to come, the hood was abuzz with the story of what had happened. Young Curtis had outmaneuvered and outsmarted the older rival. Unlike the latter, he was unafraid to do the violence himself. What he had done was bold and dramatic—it had come out of nowhere. Every time people would see Nitty on the streets with the long scar on his face, they were reminded of the incident. Rivals would now have to think twice before challenging his status—he showed he was tough and crafty. And those in his crew were duly impressed with his sangfroid and how he had turned the situation around. They now saw him differently, as somebody who could last in this jungle and was worth following.

Curtis followed this up with other similar actions, and

slowly he elevated himself above the other hustlers. Now there were younger ones who looked up to him and would soon form the core of a devoted band of disciples who would help him in his transition into music.

After the success of his first album in 2003, Curtis (now known as 50 Cent) began to realize his dream of forging a business empire. But as this took shape in the months and years to follow, he began to feel that something was wrong. It would be natural to believe that with his current position and fame, those working for him would simply follow his lead and do what he wanted. But his whole life had been a lesson in the opposite—people continually take from you; they doubt your powers and challenge you.

In this environment, his executives and managers were not trying to take his money or his life, but rather he had the feeling they were nibbling away at his power, trying to soften his image and make him more corporate and predictable. If he let this go on, he would lose the only quality that made him different—his propensity to take risks and do the unexpected. He might become a safe investment, but he would no longer be a leader and a creative force. In this world, you cannot relax and rest on your name, your past achievements, your title. You have to fight to impose your difference and compel people to follow your lead.

All of these thoughts became painfully clear to him in the summer of 2007. His third album, *Curtis*, was to be released in September of that year, and everyone seemed to be asleep. The record label, Interscope, was acting as if the album would sell itself. His management team had put together a marketing campaign that he felt was too tame, passive, and corporate. They were trying to control too much. Then one August afternoon, an employee at G-Unit Records (Fifty's own label within Interscope) told him that a video from the upcoming album somehow just got leaked to the Internet. If it spread, it would mess up the carefully orchestrated rollout of songs that had been planned for that month. Fifty was the first to hear of this, and after contemplating what to do next, he decided it was finally time to shake up the dynamic, do the unpredictable, and play the part of the hustler king.

He called into his office his radio and Internet team at G-Unit. Instead of working to contain the viral spread of the video—the usual response to such a problem and what management would advocate—he ordered them to surreptitiously leak it to other sites and let it spread like wildfire.

On top of this move, they created the following story to tell journalists for public consumption: When Fifty heard of the leak, he flew into a wild rage. He threw his phone at the window with such force he cracked it. He tore the plasma TV off the wall and smashed it into pieces. He left the building in a fit, and the last thing they heard him yelling was that he was shutting it all down and going on vacation. That evening, on Fifty's orders, they had the maintenance man for the building take pictures of the damage (all faked for this purpose), and then "leaked" the photos to the Internet. They were to keep all of this a secret—not even management was to know that this drama was completely manufactured.

In the days to come Fifty watched with satisfaction as this story spread everywhere. Interscope was awakened from its slumber. Management was sent the message that he was now in command—if he refused to do any more publicity, as he had threatened, their whole campaign was doomed. They had to follow his lead here and let him set the tone for the publicity, which meant something more aggressive and fluid. Among Fifty's executives and employees, word spread quickly of what had supposedly happened—his reputation for being unpredictable and violent brought to life. When they now saw him in the offices, they felt a twinge of fear. Better to pay attention to what he wanted than risk witnessing his anger. And for the public, this was just the kind of story they expected from the thug rapper. It compelled their attention. They could laugh at his out-of-control antics, not realizing that it was Fifty, directing the drama, who would have the last laugh.

The Fearless Approach

WHEN I REACHED THE TOP IN BUSINESS, I ADJUSTED TO MY NEW POSITION—I BECAME BOLDER AND CRAZIER THAN BEFORE. AND I LISTENED EVEN LESS TO PEOPLE WHO TRIED TO SLOW ME DOWN.

—50 Cent

Throughout history we have witnessed the following pattern: certain people stand out from the crowd because of some special skill or talent that they have. Perhaps they are masters at the political game, knowing how to charm and win the proper allies. Or maybe they have superior technical knowledge in their field. Or maybe they are the ones who initiate some bold venture that has success. In any event, these types suddenly find themselves in leadership positions, something for which their past experience and education has not prepared them.

Now they are alone and on the top, their every decision and action scrutinized by the group and the public. The pressures can be intense. And what inevitably happens is that many of them unconsciously succumb to all kinds of fears. Whereas before they might have been bold and creative, now they grow cautious and conservative, aware of the heightened stakes. Secretly scared of being held accountable for the success of the group, they over-delegate, poll everyone for their opinions, or refrain from making the hard decisions. Or they become excessively dictatorial, trying to control everything—another sign of weakness and insecurity. It is the story of great senators who make lousy presidents, bold lieutenants who turn into mediocre generals, or top-level managers who become incompetent executives.

And yet among the group there are inevitably a few who demonstrate the opposite—they rise to the position, displaying extraordinary leadership skills that no one had suspected were in them. We find in this group people like Napoleon Bonaparte, Mahatma Gandhi, and Winston Churchill. What links these people together is not some mysterious skill or bit of knowledge, but rather a quality of character, a temperament that reveals the essence of the art. They are fearless. They do not shrink from making the hard decisions by themselves—instead they seem to relish such responsibility. They do not suddenly become more conservative, but in fact show a propensity for bold action. They exhibit tremendous grace under fire.

Such types come to understand in various ways that a leader has a unique power that generally goes untapped. Any group tends to assume the spirit and energy of the person on top. If that person is weak and passive, then the group tends to splinter into factions. If such leaders lack confidence, their insecurities tend to filter their way down the line. Their nervous, fretful moods put everyone on edge. But there is always the opposite possibility. A leader who is audacious, out in front, and setting the tone and agenda for the group sparks a higher energy level and confidence. Such a person on top does not need to yell or push people around; those below want to follow his or her lead because it is strong and inspiring.

In war, where leadership skills are more immediately apparent and necessary because lives are at stake, we can distinguish two leadership styles—from behind or from the front. The former type of general likes to stay in his tent or headquarters and bark out orders, feeling that having such distance makes it easier to command. This style can also mean involving lieutenants and other generals in important decisions, choosing to lead by committee. In both cases, the commander is trying to hide himself from scrutiny, accountability, and danger. The greatest generals in history, however, are invariably those who lead from the front and by themselves. They can be seen by the troops at the head of the army, exposing themselves to the same fate as any foot soldier. The Duke of Wellington said that the mere appearance of Napoleon Bonaparte at the head of his army translated into the equivalent of an additional forty thousand men. A kind of electrical charge passes through the troops—he is sharing in their sacrifices, leading by example. It has almost religious connotations.

We notice the same two styles in business and politics as well. The executives who lead from behind will always try to disguise it as a virtue: the need for secrecy, or their desire to be more fair and democratic. But it really stems from fear and it invariably leads to a lack of respect from those below. The opposite style, leading from the front and by example, has the same power in the office as it does on the battlefield. Leaders who work harder than anyone else, who practice what they preach, who are not afraid to be accountable for tough decisions or to take risks, will find they have created a well of respect that will pay great dividends down the road. They can ask for sacrifices, punish troublemakers, and make occasional mistakes all without facing the usual grumbling and doubts. They don't have to yell, complain, and force their men and women to follow. People do so willingly.

In urban environments such as Southside Queens, respect is an extremely important issue. In other places, your background, education, or résumé might lend you some authority and credibility, but not in the hood. There, everyone starts from zero. To gain respect from your peers, you must repeatedly prove yourself. People are constantly prone to doubting your abilities and your power. You must show again and again that you have what it takes to thrive and to last. Big words and promises mean nothing; only actions carry weight. If you are authentic, as tough as you seem to be, then you will earn the respect that will make people back off and make your life that much easier.

This should be your perspective as well. You start with nothing in this world. Any titles, money, or privilege you inherit are actually hindrances. They delude you into believing you are owed respect. If you continue to impose your will because of such privileges, people will come to disdain and despise you. Instead only your actions can prove your worth. They tell people who you are. You must imagine that you are continually being challenged to show that you deserve the position you occupy. In a culture full of fakery and hype, you will stand out as someone authentic and worthy of respect.

The greatest leaders in history all inevitably learned by experience the following lesson: it is much better to be feared and respected than to be loved. As a prime example, look at the film director John Ford, the man behind some of the greatest films in Hollywood history. The task of film directors can be particularly difficult. They have to deal with large crews, actors with their delicate egos, and dictatorial producers who want to meddle every step of the way, all the while being under extreme time limits and with large amounts of money at stake. The tendency for directors is to give ground on these various battlefields—to placate and cajole the actors, to let the producers have their way here and there, to gain some cooperation by being pleasant and likable.

Ford was by nature a sensitive and empathetic man, but he learned that if he revealed this side of his personality he quickly lost control over the final product. The actors and producers would begin to assert themselves and the film would lose any sense of cohesion. He noticed that the notoriously nice directors never really lasted very long—they were pushed around and their films were lousy. Early on in his career he decided he would have to forge a kind of mask for himself—that of a man who was implacable and even a bit frightening.

On the set, he made it clear he was not the usual prima donna director. He would work longer hours than anyone. If they were filming on some location with harsh conditions, he would sleep in a tent like everyone else and share their bad food. On occasion he would get into violent fistfights on the set, most often with his leading actors, such as John Wayne. These fights were not for show; they were bruising and he engaged in them with all his strength, making the actors fight back with equal force. This would set a tone—an actor would tend to feel embarrassed by engaging in his usual prissy behavior and ego tantrums. Everyone was treated the same. Even the archduke of Austria—trying to carve out a career as a Hollywood actor—was yelled at and pushed into a ditch by Ford himself.

He had a unique way of directing actors. He would say only a few, well-chosen words about what he wanted from them. Then, if they did the wrong thing on the set, he would brutally humiliate them in front of everyone. They quickly learned they had to pay attention to the few words he spoke and to his body language on the set, which would often tell more. They had to raise their levels of concentration and bring even more of themselves into the part. Once, when the famous producer Samuel

Goldwyn visited the set, he told Ford he just wanted to watch him work (a producer's way of spying and applying pressure). Ford didn't say a word. The next day, however, he visited Goldwyn in his office and just sat silently in the chair by Goldwyn's desk, glaring at him. After a while Goldwyn, exasperated, asked him what he was doing. He just wanted to watch Goldwyn work, Ford answered. Goldwyn never visited him again on the set and quickly learned to give him his space.

All of this had a strange and paradoxical effect on the cast and crew. They came to love working for John Ford and would die to gain a place among his exclusive team of return staff. His standards were so high, it forced them to work harder—he made them superior actors and technicians. An occasional nice gesture or compliment on his part carried double the weight and would be remembered for a lifetime. The end results of his tough and unforgiving manner was that he managed to maintain a higher degree of control over the final product than most other directors, and his films were consistently of the highest quality. Nobody dared to challenge his authority and he lasted in Hollywood as the king of Westerns and action films for well over forty years—an unprecedented achievement in the industry.

Understand: to be a leader often requires making tough choices, getting people to do things against their will. If you have chosen the soft, pleasing, compliant style of leadership, out of fear of being disliked, you will find yourself with less and less room to compel people to work harder or make sacrifices. If you suddenly try to be tough, they often feel wounded and personally upset. They can move from love to hate. The opposite approach yields the opposite result. If you build a reputation for toughness and getting results, people might resent you, but you will establish a foundation of respect. You are demonstrating genuine qualities of leadership that speak to everyone. Now with time and a well-founded authority, you have room to back off and reward people, even to be nice. When you do so, it will be seen as a genuine gesture, not an attempt to get people to like you, and it will have double the effect.

Keys to Fearlessness

FOR IT IS A GENERAL RULE OF HUMAN NATURE THAT PEOPLE DESPISE THOSE WHO TREAT THEM WELL AND LOOK UP TO THOSE WHO MAKE NO CONCESSIONS.

—Thucydides

Thousands of years ago, our most primitive ancestors formed groups for power and protection. But as these groups got larger, they encountered a problem with human nature that plagues us to this day. Individuals have different levels of talent, ambition, and assertiveness; their interests do not necessarily converge on all points. When it comes to the important decisions upon which the fate of the tribe hangs, the members will often think of their own narrow agendas. A group of humans is always on the verge of splintering into a chaos of divergent interests.

For this purpose, leaders were chosen to make the hard decisions and end all the dissension. But the members of the tribe would inevitably feel ambivalence towards their leaders. They saw the necessity for them and the respect that should be paid to their authority, but they feared that their chieftains and kings would accumulate too much power and oppress them. They often wondered why this particular person or family deserved such a lofty position. In many ancient cultures, the king was ritually put to death after a few years to ensure he would not turn into an oppressor. In more advanced ancient civilizations, there were constant rebellions against those in power—much more intense and numerous than anything we have known in the modern era.

Of all the leaders in ancient times who had to deal with such difficulties, none stands out more than Moses. He had been chosen by God to lead the Hebrews out of slavery in Egypt and to the Promised Land. Although the Hebrews suffered in Egypt, they had relative security. Moses wrested them from this predictable life and set them to wander for forty years in the wilderness, where they were plagued by a lack of food, shelter, and basic comforts. They constantly doubted Moses and even came to hate him— some plotting to kill him, as the king who needed to be sacrificed. They saw him as an oppressor and madman. To aid his cause, God would perform regular miracles to show that Moses was chosen and blessed, but these miracles were quickly forgotten and the Hebrews kept resorting to their endless complaining and recalcitrance.

To overcome the seemingly impossible obstacles in his

path, Moses resorted to a unique solution: he united the twelve constantly divided tribes around a single, simple cause—one God to worship, and the attainable goal of reaching the Promised Land. He was not there for power or glory but merely to lead them to this much-desired goal. Moses could not afford to absent himself for a day or two, or to ease up on his leadership. The tribes were continually prone to doubt him and forget the larger picture, the reason for their suffering. The Hebrew word for "lead" means to be out in front, to drive. He had to be out there constantly in the vanguard, unifying them around his vision of the Promised Land. This meant being ruthless with internal dissenters, putting whole families to death that stood in the way of the larger cause.

In essence, Moses learned to play a role for the Hebrews—the man who is possessed of a vision from God, indomitable in spirit, and acting for the greater good. A normal tribe member would have to ask him- or herself if this Promised Land was not something that existed merely in Moses's mind. But the force of his conviction and the determination to lead people to the Promised Land made it hard to doubt him. He had to play this role to the hilt to convince them his top position was legitimate and sanctioned by God. His ability to lead such a fractious group for some forty years has to be considered the greatest masterpiece of leadership in history.

We moderns believe we have moved far beyond our primitive, tribal origins. After all, we live in a secular, rational world. A leader today needs to possess certain technical and managerial skills. But three thousand years of civilization have not altered human nature, and in fact the endless difficulties that plagued such leaders as Moses have become only more acute. Whereas before we humans might think first of the tribe, we now think primarily of ourselves, our careers, and our narrow interests. Office politics is the extreme endpoint of this trend.

We are now more distracted than ever, with thousands of bits of information competing for our attention in the course of a day. This makes us less patient and capable of seeing the larger picture. If we were being led out of slavery, we would not be able to focus on the Promised Land for more than a few minutes. We are much more skeptical when it comes to those in authority. We still feel the ancient ambivalence towards rulers; instead of sacrificing them, we feed them to the press and secretly gloat in their downfall. To be a leader now means overcoming these aspects of human nature while still seeming to be fair and decent—an almost impossible task.

At the same time, however, people feel this division and selfishness as a depressing phenomenon. They desperately want to believe in a cause, to work for the greater good, to follow a leader who imbues them with a sense of purpose. They are more than receptive to the kind of quasi-religious leadership that Moses embodies. As the one on top, you must rid yourself of your modern prejudices, your fetishism of technical means. To be a leader still means that you are playing a role, out in front, fearlessly driving the group forward. If you fail to unify the group around some glorious cause, some equivalent of the Promised Land, then you will find that you are having to push and pull your followers, who are constantly splitting up into factions. Instead you must assume a prophetic air, as if you were merely chosen to lead them towards some higher goal. You are compelling them to follow on their own, making less a show of personal power and more a demonstration of the cause that unites them all. This will give you the proper authority to lead and an aura of power.

To master the art of leadership you must see yourself as playing certain parts that will impress your disciples and make them more likely to follow you with the necessary enthusiasm. The following are the four main roles you must learn to perform.

THE VISIONARY

By the beginning of the twentieth century, Thomas Alva Edison was seen as America's preeminent inventor and scientist. His research labs were the source of some of the most important technological breakthroughs of the time. But the truth is that Edison himself had only a few months of formal education and was not really a scientist at all. Instead he was a mix of visionary, strategist, and shrewd businessman.

His method was simple: he scoured the globe, looking at all of the latest advances in science and technology. With his understanding of business and the latest social trends, he thought long and hard about how some of these advances could be translated into products with great commercial appeal, that could transform how people lived—electricity lighting up cities, improved telephones altering the course of commerce, motion pictures entertaining the masses. He would then hire the best minds in these fields to bring to life his ideas. Every product that came through his lab was inevitably stamped with Edison's particular vision and sense of marketing.

Understand: a group of any size must have goals and long-term objectives to function properly. But human nature serves as a great impediment to this. We are naturally consumed by immediate battles and problems; we find it very difficult, if not unnatural, to focus with any depth on the future. Thinking ahead requires a particular thought process that comes with practice. It means seeing something practical and achievable several years down the road, and mapping out how this goal can be achieved. It means thinking in branches, coming up with several paths to get there, depending on circumstances. It means being emotionally attached to this idea, so that when a thousand distractions and interruptions seem to push you off course, you have the strength and purpose to keep at it.

Without one person on top who charts the way to this larger goal, the group will wander here and there, grasping at schemes for quick money, or be moved by the narrow political aspirations of one member or another. It will never accomplish anything great. You as the leader are the only bulwark against this endless wandering. You must have the strength to stamp the group with your own personality and vision, giving it a core and an identity. If you lose sight of the larger picture, then only bad things will ensue.

You must play this visionary role with some dramatic flair, like Edison who was a consummate performer and promoter.

He would give dazzling presentations of his ideas, and stage events to get on the front page of newspapers. Like Moses describing the Promised Land, he could paint an alluring picture of the future that his inventions would help create. This drew in money from investors and inspired his researchers to work even harder. Your own level of excitement and self-belief will convince people that you know where you are going and should be followed.

THE UNIFIER

When Louis XIV began ruling France in 1661, he inherited an almost impossible situation. The feudal dukes and lords of France maintained tight control over their various realms. Recent ministers such as the Duke de Richelieu and Cardinal Mazarin had made most of the important decisions that lay outside the control of the lords. The king had been mostly a figurehead, presiding over a deeply fractured country whose power in Europe had been on the decline for quite some time.

Louis was determined to reverse all of this, and his method was powerful and dramatic. At first he kept his intentions to himself, and then suddenly he announced to one and all that he would not appoint a minister to run the country—from now on that would be his task. Next, he ordered the aristocracy to take up residence in the palace of Versailles that he had recently constructed. The closer they lived to him in the palace, the more influence they would have; if they remained in their duchies, to conspire against him, they would find themselves isolated from the new center of power he had created.

His most brilliant maneuver, however, was the most subtle one of all. He created a cause for the French people to believe in—the greatness and glory of France itself, which had as its mission to be the center of civilization and refinement, the model for all of Europe. For this purpose, he led the country into various wars to extend France's political might. He became the preeminent patron of the arts, making France the cultural envy of Europe. He created impressive spectacles to delight and distract the public from his power moves. The nobility were not fighting for Louis but for the greatness of the nation. In this way, he transformed a deeply divided, almost chaotic country into the supreme power of Europe.

Understand: the natural dynamic of any group is to splinter into factions. People want to protect and promote their narrow interests, so they form political alliances from within. If you force them to unite under your leadership, stamping out their factions, you may take control but it will come with great resentment—they will naturally suspect you are increasing your power at their expense. If you do nothing, you will find yourself surrounded by lords and dukes who will make your job impossible.

A group needs a centripetal force to give it unity and cohesion but it is not enough to have that be you and the force of your personality. Instead it should be a cause that you fearlessly embody. This could be political, ethical, or progressive—you are working to improve the lives of people in your community, for instance. This cause elevates your group above others. It has a quasi-religious aura to it, a kind of cult feeling. Now, to fight or doubt you from within is to stand against this cause and seem selfish. The group, infused with this belief system, will tend to police itself and root out troublemakers. To play this role effectively, you must be a living example of this cause, much as Louis exemplified the civilizing power of France in his own carefully crafted behavior.

THE ROLE MODEL

You cannot control a large group on your own. You will turn into a micromanager or dictator, making yourself exhausted and hated. You need to develop a team of lieutenants who are infused with your ideas, your spirit, and your values. Once you have such a team, you can give them latitude to operate on their own, learning for themselves and bringing their own creativity to the cause.

This is the system that Napoleon Bonaparte initiated and has since been imitated by the greatest generals of the modern era. He would give his field marshals a clear sense of the goals for a particular campaign or battle, what has become known as the "mission statement." They were then empowered to reach those goals on their own, in their own way. All that mattered were the results. The idea behind it is that those who are fighting on the ground often have a better sense of what needs to be done in the here and now; they have more information at their fingertips than the leader. With a degree of trust in their decisions, they can operate fast and feel more engaged in the execution of the war. This revolutionary system allowed Napoleon's army to move with greater speed and to cultivate a team of highly experienced and brilliant field marshals. And it took great courage on his part to trust in them and not try to control everything on the battlefield.

Operating with a mission statement is an effective way of softening your image and disguising the extent of your power. You are seen as more than just a leader; you are a role model, instructing, energizing, and inspiring your lieutenants. In crafting this team, look for people who share your values and are open to learning. Do not be seduced by a glittering résumé. You want them near you, to absorb your spirit and ways of doing things. Once you feel they have the proper training, you must not be afraid to let go of the reins and give them more independence. In the end, this will save you much energy and allow you to continue focusing on the greater strategic picture.

THE BOLD KNIGHT

Every group has a kind of collective energy, and on its own this will tend towards inertia. This comes from people's powerful desires to keep things comfortable, easy, and familiar. Over time, in any group, conventions and protocol will assume greater importance and govern people's behavior. The larger the group, the more conservative it will tend to become, and the greater this force of inertia. The paradox is that this defensive, passive posture has a depressing effect on morale, much like sitting in one place for too long will lower your spirits.

More than likely you rose to the top by virtue of your boldness and desperate desire to get ahead. You took risks that made you rise to the occasion with all of your energy and creativity, and this fearless spirit attracted positive attention. The group inertia will naturally tend to tamp all of that down and neutralize the source of your power.

Since you are the leader, you are the one who can alter this and set a pace that is more alive and active. You remain the bold and enterprising knight. You force yourself to initiate new projects and domains to conquer; you take proactive measures against possible dangers on the horizon; you seize the initiative against your rivals. You keep your group marching and on the offensive. This will excite them and give them a feeling of movement. You are not taking unnecessary risks, but simply adding a dash of aggression to your normally staid group. They become used to seeing you out in front and grow addicted to the excitement you bring with each new campaign.

Reversal of Perspective

We live in times of great mistrust of any form of authority. Some of this springs from envy of those who have power and have achieved something. Some of it comes from experiences with people who abuse their position of power to get their way. In any event, such distrust makes it harder and harder to be a strong and effective leader. Under the sway of this leveling force, you yourself might be tempted to act with less authority, to be more like everyone else, or to make yourself likable. This will only make your job that much harder. Instead it is better that you see the whole concept of authority in a different light.

The word "authority" comes from the Latin root *autore*, meaning author—a person who creates something new. This could be a work of art, a new way of operating in the world, or new values. The health of any society depends on those who infuse it with such innovations. These works or actions by individuals give them credibility and authority to do more. The great Roman general Scipio Africanus the Elder invented a whole new style of warfare in the campaign against Hannibal that was tremendously successful. This gave him the authority to lead the campaign itself, and later to launch a political career. For the Romans, if you simply acted as if your position entitled you to certain powers, you lost your authority. You were no longer an author, a contributor, but a passive consumer of power.

As a leader this is how you must view yourself as well. You are an author creating a new order, writing a new act in some drama. You never rest on your laurels or past achievements. Instead you are constantly taking action that moves the group forward and brings positive results; that record speaks for itself. Despite the spirit of the times, people have a secret yearning to be guided by a firm hand, by someone who knows where they are going. It is distressful to always feel distracted and wandering. The members of your group will give you the respect and authority you require if you earn it as an author and creator. In the end, if people mistrust and resist your authority, you really have only yourself to blame.

A DISTINGUISHED COMMANDER WITHOUT BOLDNESS IS UNTHINKABLE. NO MAN WHO IS NOT . . . BOLD CAN PLAY SUCH A ROLE, AND THEREFORE WE CONSIDER THIS QUALITY THE FIRST PREREQUISITE OF THE GREAT MILITARY LEADER. HOW MUCH OF THIS QUALITY REMAINS BY THE TIME HE REACHES SENIOR RANK, AFTER TRAINING AND EXPERIENCE HAVE AFFECTED AND MODIFIED IT, IS ANOTHER QUESTION. THE GREATER THE EXTENT TO WHICH IT IS RETAINED, THE GREATER THE RANGE OF HIS GENIUS.

—Carl von Clausewitz

CHAPTER 7

Know Your Environment from the Inside Out— Connection

MOST PEOPLE THINK FIRST OF WHAT THEY WANT TO EX-
PRESS OR MAKE, THEN FIND THE AUDIENCE FOR THEIR
IDEA. YOU MUST WORK THE OPPOSITE ANGLE, THINK-
ING FIRST OF THE PUBLIC. YOU NEED TO KEEP YOUR
FOCUS ON THEIR CHANGING NEEDS, THE TRENDS THAT
ARE WASHING THROUGH THEM. BEGINNING WITH THEIR
DEMAND, YOU CREATE THE APPROPRIATE SUPPLY. DO
NOT BE AFRAID OF PEOPLE'S CRITICISMS—WITHOUT
SUCH FEEDBACK YOUR WORK WILL BE TOO PERSONAL
AND DELUSIONAL. YOU MUST MAINTAIN AS CLOSE A
RELATIONSHIP TO YOUR ENVIRONMENT AS POSSIBLE,
GETTING AN INSIDE "FEEL" FOR WHAT IS HAPPENING
AROUND YOU. NEVER LOSE TOUCH WITH YOUR BASE.

Hood Economics

I KNEW THAT THE GHETTO PEOPLE KNEW THAT I
NEVER LEFT THE GHETTO IN SPIRIT, AND I NEVER
LEFT IT PHYSICALLY ANY MORE THAN I HAD TO. I HAD
A GHETTO INSTINCT; FOR INSTANCE, I COULD FEEL
IF TENSION WAS BEYOND NORMAL IN A GHETTO AU-
DIENCE. AND I COULD SPEAK AND UNDERSTAND THE
GHETTO'S LANGUAGE.

—Malcolm X

Starting out as a drug dealer at the age of twelve, Curtis Jack
son faced an unfamiliar world that contained all kinds of dangers. The business side of hustling was relatively easy to figure
out. It was the people, the various actors in the game—the
rival hustlers, the big-time dealers, the police—who could
be tricky. But strangest and most impenetrable of all was the
world of the drug users themselves, the clientele upon which

his business depended. Their behavior could be erratic and even downright frightening.

With rival hustlers and the police, Curtis could get inside their way of thinking because they all operated with a degree of rationality. But the drug fiends seemed to be dominated by their needs, and they could turn unfriendly or violent at any moment. Many dealers developed a kind of phobia of the fiends. They saw in them the weaknesses and dependence that could befall anyone who succumbed to addiction. The hustler relies on his razor-sharp mind; to even flirt with drug use could destroy such power and lead him down the slippery slope towards dependence. If he was around the fiends too much he could become a user himself. Curtis understood this and kept his distance from them, but this aspect of hustling bothered him.

On one particular occasion, the fiends were suddenly avoiding him and he could not figure out why. All he knew was that he could not sell a batch of drugs that he had on consignment. Under such an arrangement, a higher-up source, or connect, had given him the drugs for free; once he sold the entire lot, he would return a specified amount of the earnings to the connect and keep the rest as profit. But in this instance it looked like he would not make nearly enough to pay back the connect. That could prove damaging to his reputation and lead to all kinds of trouble; he might have to steal to get the money.

Feeling somewhat desperate, he went into full hustling mode, working night and day, offering all kinds of discounts, whatever it took to unload the drugs. He managed to make back just enough, but it was a close call. Perhaps the quality of the batch he was selling was inferior, but how could he tell beforehand and how could he prevent this from happening again and again?

One day he sought the advice of a man named Dre, an older hustler who had lasted an unusually long time dealing drugs on the streets. He was considered a sharp businessman (in prison, he had studied economics on his own), and he seemed to have an especially good rapport with the fiends. Dre explained to Curtis that in his experience there are two kinds of hustlers in this world—those who stay on the outside, and those who move to the inside. The outside types never bother to learn anything about their customers. It's just about money and numbers. They have no concept of psychology or the nuances of people's needs and demands. They're afraid of getting too close to the customer—that might force them to reassess their ideas and methods. The superior hustler moves to the inside. He's not afraid of the fiends; he wants to find out what's going on in their heads. Drug users are no different from anyone else. They have phobias and bouts of boredom and a whole inner life. Because you remain on the outside, he told Curtis, you don't see any of this and your hustling is purely mechanical and dead.

To raise your game, he explained, you have to first put into practice one of the oldest hustling tricks in the book— the "tester." What this means is the following: whenever you get a batch of drugs, you separate a portion of it to give out for free to certain fiends. They tell you right there on the spot whether the stuff is good or bad. If their feedback is positive, they will spread the word through their own networks, and such reports are so much more credible, coming from a fellow user, than reports from a hustler hyping his own stuff. If the feedback is negative, you will have to adjust and find some way to cut it, to offer "illusions" (apparent two for one deals, with the capsules simply loaded with dust), whatever it takes to unload it. But you must always operate with feedback on the quality of your product. Otherwise you will not survive on these streets.

Once you have this system in place, you use it to cultivate relationships with your most reliable fiends. They supply you with valuable information about any kind of change in tastes that are happening. Talking to them you get all kinds of ideas for marketing schemes and new angles for hustling. You gain a feel for how they think. From this inside position, the whole game explodes into something creative and alive with possibility.

Curtis quickly incorporated this system and soon discovered that the drug fiends were not at all as he had imagined. They became erratic only when you were not consistent in your dealings with them. They valued convenience and fast transactions, wanted something new every now and then, and loved the thought of any kind of deal. With this growing body of knowledge he could play to their needs and manipulate their demand. He discovered something else—spending much of their time on the streets, they were a great source of information about what was going on with the police, or the weaknesses of rival hustlers. Knowing so much about the neighborhood gave him a feeling of great power. Later he would translate this same strategy to music and his mix-tape campaign on the streets of New York. Maintaining a close connection to the tastes of his fans, he would alter his music to their responses and create the kind of sound that had a visceral appeal, something they had never heard before.

After the remarkable success of his first two commercial albums, Curtis (now known as 50 Cent) stood on top of the music world, but his sense of connection, so vital on the streets, was fading in this new environment he now inhabited. He was surrounded by flatterers who wanted to be in his entourage, and managers and industry people who saw in him only dollar signs. His main interactions were with people in the corporate world or other stars. He could no longer hang out on the streets or get firsthand looks at the trends that were just starting up. All of this meant that he was flying blind with his music, not really sure if it would connect anymore with his audience. They were the source of his energy and spirit, but the distance separating them was growing. Other stars seemed to not mind this; in fact, they enjoyed living in this kind of celebrity bubble. They were afraid of coming back down to earth. Fifty felt the opposite, but there seemed to be no way out.

Then in early 2007, he decided to start up his own website. He thought of it as a way to market his music and merchandise directly to the public, without the screen of his record label, which was proving quite inept in adapting to the Internet age. Soon this website transformed itself into a social networking site, like Facebook for his fans, and the more he delved into it, the more he began to sense that this represented much more than a marketing gimmick—it was perhaps the ultimate tool for reconnecting with his audience.

First, he decided to experiment. As he prepared to launch a G-Unit record in the summer of 2008, he leaked one of the songs onto the website on a Friday night, then the next day he refreshed the Comments page every few minutes and tracked the members' responses to it. After several hundred comments it was clear that the verdict was negative. The song was too soft, they judged; they wanted and expected something harder from a G-Unit record. Taking their criticisms to heart, he shelved the song and soon released another, creating the hard sound they had demanded. This time the response was overwhelmingly positive.

This called for more experiments. He put up the latest single from his archenemy, The Game, hoping to read negative comments from his fans. To his surprise, many of them liked the song. He engaged in an online debate with them and had his eyes opened about changes in people's tastes and why they might have grown distant from his music. It forced him to rethink his own direction.

To draw more people to his site, he decided to break down the distance in both directions. He posted blogs on personal subjects, and then responded to his fans' comments. They could feel they had complete access to him. Using the latest advances in phone technology, he took this further, having his team film him on their cell phones wherever he went; these images were then streamed live on the website. This generated intense traffic and online chatter—fans would never know when such moments could happen, so they were forced to check in at regular intervals to try to catch these spontaneous moments, sometimes riveting in their banality, other times made dramatic by Fifty's flair for confrontation. Membership grew by leaps and bounds.

As it evolved, the website came to strangely resemble the world of hustling that he had created for himself on the streets of Southside Queens. He could produce testers (trial songs) for his fans, who were like drug fiends, constantly hungry for new product from Fifty, and he could get instant feedback on their quality. He could develop a feel for what they were looking for and how he could manipulate their demand. He had moved from the outside to the inside, and the hustling game came alive once more, this time on a global scale.

The Fearless Approach

THE PUBLIC IS NEVER WRONG. WHEN PEOPLE DON'T RESPOND TO WHAT YOU DO, THEY'RE TELLING YOU SOMETHING LOUD AND CLEAR. YOU'RE JUST NOT LIS-TENING.

—50 Cent

All living creatures depend for their survival on their relationship to their environment. If they are particularly sensitive to any kind of change—a danger or an opportunity—they have greater power to dominate their surroundings. It is not simply that the hawk can see farther than any other creature, but that it can see great detail, picking out the slightest alteration in the landscape. Its eyes give it tremendous sensitivity and supreme hunting prowess.

We live in an environment that is mostly human. It consists of the people that we interact with day in and day out. These humans come from many varied backgrounds and cultures. They are individuals with their own unique experiences. To know people well—their differences, their nuances, their emotional life—would give us a great sense of connection and power. We would know how to reach them, communicate more effectively, and influence their actions. But so often we remain on the outside and lack this power. To connect to the environment in this way would mean having to move outside ourselves, train our eyes on people, but so often we prefer to live in our heads, amid our own thoughts and dreams. We strive to make everything in the world familiar and simple. We grow insensitive to people's differences, to the details that make them individuals.

At the root of this turning inward and disconnect is a great fear—one of the most primal known to man, and perhaps the least understood. In the beginning, our primitive ancestors formed groups for protection. To create a sense of cohesion, they established all kinds of codes of behavior, taboos, and shared rituals. They also created myths in which their tribe was considered to be the favorite of the gods, chosen for some great purpose. To be a member of the tribe was to be cleansed by rituals and to be favored by the gods. Those who belonged to other groups had unfamiliar rituals and belief systems— their own gods and origin myths. They were not clean. They represented the Other—something dark, threatening, and a challenge to the tribe's sense of superiority.

This was part of our psychological makeup for thousands upon thousands of years. It transformed itself into a great fear of other cultures and ways of thinking—for Christians, this meant all heathens. And despite millennia of civilization, it lives on within us to this day, in the form of a mental process in which we divide the world into what is familiar and unfamiliar, clean and unclean. We develop certain ideas and values; we socialize with those who share those values, who form part of our inner circle, our clique. We form factions of rigid beliefs—on the right, on the left, for this or for that. We live in our heads, with the same thoughts and ideas over and over, cocooned from the outside world.

When we are confronted with people or individuals who have different values and belief systems, we feel threatened. Our first move is not to understand them but to demonize them—that shadowy Other. Alternatively, we may choose to look at them through the prism of our own values and assume they share them. We mentally convert the Other into something familiar— "they may come from a completely different culture, but after all, they must want the same things we do." This is a failure of our minds to move outward and understand, to be sensitive to nuance. Everything must be white or black, clean or unclean.

Understand: the opposite approach is the way to power in this world. It begins with a fundamental fearlessness—you do not feel afraid or affronted by people who have different ways of thinking or acting. You do not feel superior to those on the outside. In fact, you are excited by such diversity. Your first move is to open up your spirit to these differences, to understand what makes the Other tick, to gain a feel for people's inner lives, how they see the world. In this way, you continually expose yourself to wider and wider circles of people, building connections to these various networks. The source of your power is your sensitivity and closeness to this social environment. You can detect trends and changes in people's tastes well before anyone else.

In the hood, conditions are more crowded than elsewhere; people with all kinds of different psychologies are constantly in your face. Any power you have depends on your ability to know everything that is going on around you, to be sensitive to changes, aware of the power structures that are imposed from without and within. There is no time or room to escape to some inner dreamland. You have a sense of urgency to stay connected to the environment and the people around you—your life depends on it.

We now live in similar conditions—all kinds of people of divergent cultures and psychologies are thrown together. But because we live in a society of more apparent abundance and ease, we lack that sense of urgency to connect to other people. This is dangerous. In such a melting pot as the modern world, with people's tastes changing at a faster pace than ever before, our success depends on our ability to move outside ourselves and connect to other social networks. At all costs, you need to continually force yourself outward. You must reach a point where any sense of losing this connection to your environment translates into a feeling of vulnerability and peril.

In the end this primal fear of ours translates into a mental infirmity—the closing of the mind to any ideas that are new and unfamiliar. The fearless types in history learn to develop the opposite: an open spirit, a mind that is constantly learning from experience. Look at the example of the great British primatologist Jane Goodall, whose field research revolutionized our ideas on chimpanzees and primates.

Prior to Goodall's work, scientists had established certain accepted ideas on how to do research on animals such as chimpanzees. They were mostly to be studied in cages under very controlled circumstances. On occasion, primatologists would research them in the wild; they would come up with various tricks to lure the chimpanzees closer to them, while remaining hidden behind some kind of protective screen. They would conduct experiments by manipulating the animals and noting their responses. The goal was to come up with general truths about chimpanzee behavior. Only by keeping their distance from the animals could the scientists study them.

Goodall did not have any formal training in the sciences when she arrived in 1960 in what is now known as Tanzania to study chimpanzees in the wild. Operating totally on her own, she devised a radically different means of research. The chimps lived in the remotest parts of the country and were notoriously shy. She tracked them from a distance, patiently working to gain their trust. She dressed inconspicuously and was careful to not look them in the eye. When she noticed they were uncomfortable with her being in the area, she moved away, or acted like a baboon that was merely there digging for insects.

Slowly, over the course of several months, she was able to move closer and closer. Now she could begin to identify individual chimps that she kept seeing; she gave them names, something scientists had never done before—they had always been designated by numbers. With these names, she could begin to detect subtle nuances in their individual behavior; they had different personalities, like humans. After nearly a year of this patient seduction, the chimps began to relax in her presence and allow her to interact with them, something no one had ever achieved before in the history of studying primates in the wild.

This took a tremendous degree of courage, as chimpanzees were considered the most volatile of the primates, more dangerous and violent than gorillas. As she interacted with them more and more, she noticed a change in herself as well. "I think my mind works like a chimp's, subconsciously," she wrote a friend. She felt this because she had developed an uncanny ability to find them in the forest.

Now, gaining access to them, she took note of several phenomena that belied the accepted data on chimpanzee behavior. Scientists had catalogued the animals as vegetarians; she observed them hunting and eating monkeys. Only humans were considered capable of making and using tools; she saw them crafting elaborate instruments to catch insects for food. She saw them engage in bizarre dance rituals during a rainstorm. She later observed a horrific war that went on for four years between rival packs. She catalogued some rather strange Machiavellian behavior among the males who fought for supremacy. All in all, she revealed a degree of variety in their emotional and intellectual lives that altered the concept not only of chimpanzees but also of all primates and mammals.

This has great application beyond the realms of science. Normally when you study something, you begin with certain preconceived notions about the subject. (Because scientists had come to believe that chimpanzees had a limited range of behavior, that is all that they saw, missing the much more complex reality.) Your mind begins the process in a closed state—not really sensitive to difference and nuance. You are afraid of having your assumptions challenged. Instead, like Goodall, you must let go of this need to control and narrow your field of vision. When you study an individual or a group, your goal is to get inside their minds, their experiences, their way of looking at things. To do this, you must interact with them on a more equal plane. With this open and fearless spirit, you will discover things no one had suspected before. You will have a much deeper appreciation for the targets of your actions or the public you are trying to reach. And with such understanding will come the power to move them.

Keys to Fearlessness

FEW PEOPLE HAVE THE WISDOM TO PREFER THE CRITI-CISM THAT WOULD DO THEM GOOD, TO THE PRAISE THAT DECEIVES THEM.

—François de La Rochefoucauld

In the work that we produce for business or for culture, there is always a telling moment—when it leaves our hands and reaches the public for which it was intended. In that instant it ceases to be something that was in our heads; it becomes an object that is judged by others. Sometimes this object connects with people in a profound way. It strikes an emotional chord, resonates, and has warmth. It meets a need. Other times it leaves people surprisingly cold—in our minds we had imagined it having a much different effect.

This process can seem rather mysterious. Some people seem to have a knack for creating things that resonate with an audience. They are great artists, politicians with the popular touch, or business people who are endlessly inventive. Sometimes we ourselves produce something that works, but we fail to understand why, and lacking this knowledge, we cannot reproduce our success.

There is an aspect to this phenomenon, however, that is explicable. Anything we create or produce is for a public— large or small, depending on what we do. If we are the type that lives mostly in our heads, imagining what the intended public will like, or not even caring, this spirit is reproduced in the work itself. It is disconnected from the social environment; it is a product of a person who is wrapped up in him- or herself. If, on the other hand, we are deeply connected to the public, if we have a profound sense of their needs and wants, then what we make tends to resonate. We have internalized the way of thinking and feeling of our audience and it shows in the work.

The great Russian writer Fyodor Dostoyevsky had almost two separate parts to his career: in the first, he was a socialist who interacted mostly with other intellectuals. His novels and stories were relatively successful. But then in 1849 he was sentenced to several years of prison and hard labor in Siberia for ostensibly conspiring against the government. There, he suddenly discovered that he hadn't known the Russian people at all. In prison he was thrown in among the dregs of society. In the small village where he did his hard labor, he finally mingled with the Russian peasantry that dominated the country. Once he was freed, all of these experiences became deeply embedded in his work, and suddenly his novels resonated far beyond intellectual circles. He understood his public, the mass of Russian people, from the inside, and his work became immensely popular.

Understand: you cannot disguise your attitude towards the public. If you feel superior at all, part of some chosen elite, then this seeps out in the work. It is conveyed in the tone and mood. It feels patronizing. If you have little access to the public you are trying to reach but you feel that the ideas in your head cannot fail to be interesting, then it almost inevitably comes across as something too personal, the product of someone who is alienated. In either case, what is really dominating the spirit of your work is fear. To interact closely with the public and get its feedback might mean having to adjust your "brilliant" ideas, your preconceived notions. This might challenge your tidy vision of the world. You might disguise this with a snobbish veneer, but it is the age-old fear of the Other.

We are social creatures who make things in order to communicate and connect with those around us. Your goal must be to break down the distance between you and your audience, the base of your support in life. Some of this distance is mental—it comes from your ego and the need to feel superior. Some of it is physical—the nature of your business tends to shut you off from the public with layers of bureaucracy. In any event, what you are seeking is maximum interaction, allowing you to get a feel for people from the inside. You come to thrive off their feedback and criticism. Operating this way, what you produce will not fail to resonate because it will come from the inside. This deep level of interaction is the source of the most powerful and popular works in culture and business, and a political style that truly connects.

The following are four strategies you can use to bring yourself closer to this ideal.

CRUSH ALL DISTANCE

The French artist Henri de Toulouse-Lautrec came from one of the oldest aristocratic bloodlines in France, but from early on he felt estranged from his family. Part of this came from his physical handicap—his legs had stopped growing at the age of fourteen, giving him a dwarfish appearance. Part of it came from his sensitive nature. He turned to painting as his only interest in life, and in 1882, at the age of eighteen, he moved to Paris to study with a famous artist whose studio was in Montmartre—the bohemian and somewhat seedy part of the city. There Toulouse-Lautrec discovered a whole new world— the cafés and dance halls frequented by prostitutes, con artists, dancers, street performers, and all the shady characters who found themselves drawn to this *quartier*. Perhaps because of his own alienation from his family, he identified with these outcasts. And slowly he began to immerse himself deeper and deeper in the social life of Montmartre.

He befriended the prostitutes and hired them as models, seeking to capture the essence of their lives on canvas. He returned to the dance halls often and sketched while he watched. He drank with the criminal types and the anarchist agitators who passed through the neighborhood. He absorbed every aspect of this world, including the habits of the rich people who came to the area for entertainment and to slum it. Other painters like Degas and Renoir, who both lived in Montmartre, painted many scenes of life there, but it was always with a sense of distance, as if they were outsiders peeking in. Toulouse-Lautrec was more of an active participant. And as his drawings and paintings began to reflect this immersion, his work drew more attention from the public.

All of this culminated in the posters that he did for the dance hall the Moulin Rouge, which opened in 1889. The first and most famous one of all was a scandalous image of a dancer kicking so high you can see her underwear. The colors are intense and garish. But strangest of all is the kind of flat space he created, which gives viewers the sensation that they are there onstage with the performers, in the middle of all the activity and bright lights. No one had created anything quite like it before. When the poster was placed all over the city, people were mesmerized by the image. It seemed to vibrate with a life of its own. More and more posters followed of all the figures in the Moulin Rouge whom he came to know on intimate terms, and an entire new aesthetic was forged around his complete, democratic mingling with his subjects. His work became immensely popular.

Understand: in this day and age, to reach people you must have access to their inner lives—their frustrations, aspirations, resentments. To do so, you must crush as much distance as possible between you and your audience. You enter their spirit and absorb it from within. Their way of looking at things becomes yours, and when you re-create it in some form of work, it has life. What shocks and excites you will then have the same effect on them. This requires a degree of fearlessness and an open spirit. You are not afraid to have your whole personality shaped by these intense interactions. You assume a radical equality with the public, giving voice to people's ideas and desires. What you produce will naturally connect, in a deep way.

OPEN INFORMAL CHANNELS
OF CRITICISM AND FEEDBACK

When Eleanor Roosevelt entered the White House as the First Lady in 1933, it was with much trepidation. She had a disdain for conventional politics and for the kind of cliquish attitude it fostered. In her mind, her husband's power would depend on his connection to the people who had elected him. To get out of the Depression, the public had to feel engaged in the struggle, not merely be seduced by speeches and programs. When people feel involved they bring their own ideas and energy to the cause. Her fear was that the bureaucratic nature of government would swallow up her husband. He would come to listen to his cabinet members and experts; his contact with the public would be relegated to formal channels such as reports, polls, and studies. This isolation would spell his doom, cutting him off from his base of support. Denied an official position within the administration, she decided to work to create informal channels to the public on her own.

She traveled all over the country—to inner cities and remote rural towns—listening to people's complaints and needs. She brought many of these people back to meet the president to give him firsthand impressions of the effects of the New Deal. She started a column in *The Woman's Home Companion*, in which she had posted above the headline, "I want you to write me." She would use her column as a kind of discussion forum with the American public, encouraging people to share their criticisms. Within six months she had received over 300,000 letters, and with her staff she worked to answer every last one of them. She opened other channels of communication, for instance, planting her aides in various New Deal programs who would then poll on her behalf the public affected by these programs.

With this system in place, she began to see a pattern from the bottom up—a growing disenchantment with the New Deal. Every day, she left a memo in her husband's basket, reminding him of these criticisms and the need to be more responsive. And slowly she began to have an influence on his policy, pushing him leftward—for instance, getting him to create programs such as the NYA, the National Youth Administration, which would involve young people actively in the New Deal. Over time she became the unofficial channel of communication for women's groups and African Americans, shoring up FDR's support in these two key constituencies. All of this work took tremendous courage, for she was continually ridiculed for her activist approach, long before any first lady had ever thought of taking such a role. And her work played a major part in FDR's ability to maintain his image as a man of the people.

As Eleanor understood, any kind of group tends to close itself off from the outside world. It is easier to operate this way. From within this bubble, people will delude themselves into thinking they have insight into how their audience or public feels—they read the papers, various reports, the poll numbers, etc. But all of this information tends to be flat and highly filtered. It is much different when you interact directly with the public and hear in the flesh their criticisms and feedback. You discover what lies at the root of their discontent, the various nuances of how your work affects them. Their problems come to life, and any solutions you come up with have more relevance. You create a back-and-forth dynamic in which their ideas, involvement, and energy can be harnessed for your purposes. If some distance between you and the public must be maintained, by the nature of your group or enterprise, then the ideal is to open up as many informal channels as possible, getting your feedback straight from the source.

RECONNECT WITH YOUR BASE

We see the following occur over and over: a person has success when they are younger because they have deep ties with a social group. What they produce and say comes from a real place and connects with an audience. Then slowly they lose this connection. Success creates distance. They come to spend most of their time with other successful people. Consciously or unconsciously, they come to feel separated and above their audience. The intensity in their work is gone and with it any kind of real effect on the public.

In his own way the famous black activist Malcolm X struggled with this problem. He had spent his youth as a savvy street hustler, ending up in prison on drug charges. There, he discovered the religion of Islam, as practiced by the Nation of Islam, and immediately converted. Out of prison he became a highly visible spokesperson for the group. Eventually he broke off from the Nation of Islam and transformed himself into a leading figure in the growing black power movement of the 1960s.

In these various phases of his life, Malcolm felt intense anger and frustration at the levels of injustice for African Americans, much of which he had experienced firsthand. He channeled these emotions into powerful speeches, seeming to give voice to the anger that many felt who lived deep within the ghettos of America. But as he became more and more famous, he felt some anxiety. Other leaders in the black community that he had known had begun to live fairly well; they could not help but feel some distance and superiority to those they were supposed to represent—like a father caring for a child.

Malcolm hated that feeling of creeping paternalism. In his mind, people can only help themselves—his role was to inspire them to action, not act in their name. To inoculate himself against this psychic distance, he increased his interactions with street hustlers and agitators, the kind of people from the lower depths that most leaders would scrupulously avoid. Those from the heart of the ghetto were his power base and he had to reconnect with them. He made himself spend more time with those who had suffered recent injustices, soaking up their experiences and sense of outrage. Most people mellow with age—he would retain his anger, the intensity of emotions that had impelled him in the first place and given him his charisma.

The goal in connecting to the public is not to please everyone or to spread yourself out to the widest possible audience. Communication is a power of intensity, not extensity and numbers. In trying to widen your appeal, you will substitute quantity for quality and you will pay a price. You have a base of power—a group of people, small or large, which identifies with you. This base is also mental—ideas you had when you were younger, which were tied to powerful emotions and inspired you to take a particular path. Time and success tend to diffuse the sense of connection you have to this physical and mental base. You will drift and your powers of communication will diminish. Know your base and work to reconnect with it. Keep your associations with it alive, intense, and present. Return to your origins—the source of all inspiration and power.

CREATE THE SOCIAL MIRROR

Alone, in our minds, we can imagine we have all kinds of powers and abilities. Our egos can inflate to any size. But when we produce something that fails to have the expected impact, we are suddenly faced with a limit—we are not as brilliant or skilled as we had imagined. In such a case, our tendency is to blame others for not understanding it or getting in our way. Our egos are bruised and delicate—criticism from the outside seems like a personal attack, which we cannot endure. We tend to close ourselves off and this makes it doubly difficult to succeed with our next venture.

Instead of turning inward, consider people's coolness to your idea and their criticisms as a kind of mirror they are holding up to you. A physical mirror turns you into an object; you can see yourself as others see you. Your ego cannot protect you—the mirror does not lie. You use it to correct your appearance and avoid ridicule. The opinions of other people serve a similar function. You view your work from inside your mind, encrusted with all kinds of desires and fears. They see it as an object; they see it as it is. Through their criticisms you can get closer to this objective version and gradually improve what you do. (One caveat: beware of feedback from friends whose judgments could be tainted by feelings of envy or the need to flatter.)

When your work does not communicate with others, consider it your own fault—you did not make your ideas clear enough and you failed to connect with your audience emotionally. This will spare you any bitterness or anger that might come from people's critiques. You are simply perfecting your work through the social mirror.

Reversal of Perspective

Science and the scientific method are very powerful and practical pursuits of knowledge that have come to dominate much of our thinking for the past few centuries. But they have also spawned a peculiar preconception—that to understand anything we must study it from a distance and with a detached perspective. For example, we tend to judge a book that is full of statistics and quotes from various studies as carrying more weight because it seems to have that requisite scientific objectivity and distance. Science, however, often deals with matter that is inorganic or has a marginal emotional life. Studying such things from a detached perspective makes sense and yields profound results. But this does not translate so well when dealing with people and creatures who respond from an emotional core. The knowledge of what makes them tick on the inside is missing. To study them from the outside is merely a prejudice, often one stemming from fear—dealing with people's experiences and subjectivity is messy and chaotic. Distance is cleaner and easier.

It is time to reevaluate this preconception and see things from the opposite perspective. Knowledge of human nature and social factors, the kind that is often most valuable to us, depends on knowing people and networks from the inside, on getting a feel for what they are experiencing. This can best be gained by an intense involvement and participation, as opposed to the pseudoscientific pose of the intellectual addicted to studies, citations, and numbers, all designed to back up their preconceptions. This other form of knowledge, from the inside, must be the one that you come to esteem above all others in social matters. It is what will give you power to affect people. To the extent that you feel yourself to be distant and on the outside, you must tell yourself you do not understand what you are studying or trying to reach—you are missing the mark and there is work to be done.

A REALLY INTELLIGENT MAN FEELS WHAT OTHER MEN ONLY KNOW.

—Baron de Montesquieu

CHAPTER 8

Respect the Process— Mastery

THE FOOLS IN LIFE WANT THINGS FAST AND EASY—
MONEY, SUCCESS, ATTENTION. BOREDOM IS THEIR
GREAT ENEMY AND FEAR. WHATEVER THEY MANAGE
TO GET SLIPS THROUGH THEIR HANDS AS FAST AS IT
COMES IN. YOU, ON THE OTHER HAND, WANT TO OUT-
LAST YOUR RIVALS. YOU ARE BUILDING THE FOUN-
DATION FOR SOMETHING THAT CAN CONTINUE TO
EXPAND. TO MAKE THIS HAPPEN, YOU WILL HAVE TO
SERVE AN APPRENTICESHIP. YOU MUST LEARN EARLY
ON TO ENDURE THE HOURS OF PRACTICE AND DRUDG-
ERY, KNOWING THAT IN THE END ALL OF THAT TIME WILL
TRANSLATE INTO A HIGHER PLEASURE—MASTERY OF
A CRAFT AND OF YOURSELF. YOUR GOAL IS TO REACH
THE ULTIMATE SKILL LEVEL—AN INTUITIVE FEEL FOR
WHAT MUST COME NEXT.

————————————————————————————

Slow Money

————————————————————————————

MASTER THE INSTRUMENT, MASTER THE MUSIC, THEN FORGET ALL THAT SHIT AND PLAY.

—Charlie Parker

Growing up in Southside Queens, the only people Curtis Jackson could see who had any money and power were the street hustlers. So at the age of eleven and with big dreams for the future, he chose just such a path for himself. Almost immediately, however, he saw that the life of a hustler was not glamorous at all. It consisted mostly of standing on a street corner day after day, selling the same stuff to the same fiends. It meant enduring hours with nothing to do, waiting for customers to come by, often in the bitter cold or the blistering heat. And in those long, tedious hours on the streets, Curtis's mind naturally would wander; he would find himself wishing for money that would come faster and easier, with more excitement. There were opportunities for this in the hood—they mostly involved crime or some dubious scheme. Sometimes he would feel tempted to try them, but in such moments he would remind himself of the endless stories of the hustlers he had known who had fallen for the illusion of fast, easy money—all suckers who inevitably ended up dead or broke.

There was his friend TC who got tired of hustling and fell in with a crew that would spend the summer robbing convenience stores and occasionally a bank. He made quite a haul of money over those three months and then blew it all over the fall and winter. The following summer he was back at it again. It wasn't just the money; it was the thrill that came in flirting with so much danger. But that second summer his luck ran out and he was killed in a gunfight with the police.

There was Curtis's colleague Spite, a few years older, who had managed to save some money from his hustling but had dreams of something much bigger. He convinced himself he could make a fortune fast by buying a piece of a franchise business that was new to the hood but that he felt was certain to be hot. He poured all of his money into the venture, but he was too impatient. He had not taken the time to accustom the public to his new life. Everyone believed the business was just a front for some drug operation. They avoided it and it soon became a hangout for hustlers and fiends. It failed within a few months and he never recovered from the experience.

This was the gist of the problem: to be a successful hustler you had to accustom yourself to the slow, grinding pace of the job. But in the hood, the future rarely seemed promising. It was hard for hustlers to imagine saving their money for some rainy day in the future when that day would probably never come. Inevitably the desire for something faster would creep into their blood, and if they gave in, they created a cycle they could never escape. If they were able to get some fast money, it would act like a drug—they would get excited and spend it all on items to impress people. With no money left, they would return to dealing drugs, but now it seemed too slow and boring. They would try again for something fast. They became trapped by their own greed, and as the years went by, they would fail to develop any kind of patience or discipline. They could not manage this up-and-down pace for too long. By the age of twenty-five or thirty they would burn themselves out and have no skills or money to show for their years of work. Their fate after that was generally unpleasant.

To resist this temptation, Curtis decided he would force himself in the opposite direction. He treated hustling as a job. He showed up on the street corner at the same hour every day, working from dawn to dusk. Gradually he accustomed himself to this slow pace. During the long hours with nothing to do, he would contemplate the future and come up with detailed plans of what he would accomplish year by year— ending with his eventual escape from street hustling. He would move into music, and then into business. To take the first step, he would have to save his money. The thought of this goal helped him endure the daily tedium of the job. In these slow hours, he also devised new hustling schemes, with the idea of continually improving himself at this job.

He took up boxing to discipline his mind and body. He was terrible at first, but he was tenacious, training day in and day out, eventually becoming a skilled fighter. This taught him invaluable lessons—he could get whatever he wanted through sheer persistence rather than by violence or force; progressing step-by-step was the only way to succeed in anything. By the age of twenty, he made his break into music—all according to his original plan.

In 1999, after a few years of apprenticing with Jam Master Jay, Curtis (now known as 50 Cent) signed a deal with Columbia Records. It seemed like a dream come true, but as he looked around at the other rappers who had been at the label a little longer, he saw the dangers around him had only increased. The tendency, as he saw it, was to immediately let up in your energy and focus. Rappers would feel that they had arrived, and unconsciously they wouldn't work as hard and would spend less time learning their craft. That sudden influx of money would go to their heads; they would imagine they had the golden touch and could keep it coming. One hit song or record would make this even worse. Not building something slowly—a career, a future—it would all fall apart within a few years, as younger and more eager rappers would take their place. Their life would be all the more miserable for having once tasted some glory.

To Curtis, the solution was simple: this was a new world he had entered. He had to take his time and learn it well. In the fast environment of hip-hop, he would slow everything down. He avoided the partying and kept mostly to himself. He decided to treat Columbia Records as a university, his one chance to educate himself in the business. He would record his music at night and spend the entire day at the Columbia offices, talking shop to people in every division. Gradually he taught himself more and more about marketing and distribution, and the nuts and bolts of the business. He studied all aspects of production, what went into making a hit song. He practiced his music over and over. When the label sent him and dozens of other rappers to a retreat in upstate New York to write songs, he returned with thirty-six tracks, while most of the others could barely muster five or six.

In the wake of the assassination attempt on Fifty in 2000, Columbia Records dropped him from the label, but by then he had outgrown his need for their expertise. He had accumulated so much knowledge and skill that he was able to apply it all to his mix-tape campaign, creating songs at an insane pace and marketing his music as smartly as any professional. Step by step he advanced, the campaign gaining the attention of Eminem, who signed him to his label at Interscope in 2003.

Years later he found himself in the corporate world, and he quickly discovered it was not much different from the streets. So many of the business people and executives he met had that same level of impatience. They could only think in terms of months or weeks. Their relationship to money was emotional—a way to impose their importance and feed their ego. They would come to him with schemes that seemed intriguing in the present but that led nowhere down the road. They were not attuned to the immense changes going on in the world and planning to exploit them in the future—that would take too much effort and time.

These business types came at him from all directions with endorsement deals that would make him some fast millions. They assumed he was like all the other rappers who grabbed at such opportunities. But endorsement deals would not help him build anything solid or real. It was illusion money. He would turn them down, opting to start his own businesses on his own terms—each business building on the other like links on a chain. The goal this time was simple—to forge an empire that would last. And as before, he would get there through his own grinding persistence.

The Fearless Approach

MOST PEOPLE CAN'T HANDLE BOREDOM. THAT MEANS THEY CAN'T STAY ON ONE THING UNTIL THEY GET GOOD AT IT. AND THEY WONDER WHY THEY'RE UNHAPPY.

—50 Cent

For our most primitive ancestors, life was a constant struggle, entailing endless labor to secure food and shelter. If there was any free time, it generally was reserved for rituals that would give meaning to such a hard life. Then, over thousands of years of civilization, life gradually became easier for many, and with that came more and more free time. In such moments, there was no need to work the fields or worry about enemies or the elements—just an expanse of hours to somehow fill. And suddenly a new emotion was born into this world—boredom.

At work or in rituals, the mind would be filled with various tasks to accomplish; but alone in one's house, this free time would allow the mind to roam wherever it wanted. Confronted with such freedom, the mind has a tendency to gravitate towards anxieties about the future—possible problems and dangers. Such empty time faintly echoes the eternal emptiness of death itself. And so with this new emotion that assailed our ancestors came a desire that haunts us to this day—to escape boredom at all cost, to distract ourselves from these anxieties.

The principal means of distraction are all forms of public entertainment, drugs and alcohol, and social activities. But such distractions have a drug-like effect—they wear off. We crave new ones, faster ones, to lift us out of ourselves and divert us from the harsh realities of life and creeping boredom. An entire civilization—ancient Rome—practically collapsed under the weight of this new need and emotion. Their economy became tied to the creation of novel luxuries and entertainments that sapped its citizens' spirit; few were willing anymore to sacrifice their pleasures for hard work or the public good.

This is the pattern that boredom has created for the human animal ever since: we look outside ourselves for diversions and grow dependent on them. These entertainments have a faster pace than the time we spend at work. Work then is experienced as something boring—slow and repetitive. Anything challenging, requiring effort, is viewed the same way—it's not fun; it's not fast. If we go far enough in this direction, we find it increasingly difficult to muster the patience to endure the hard work that is required for mastering any kind of craft. It becomes harder to spend time alone. Life becomes divided between what is necessary (time at work) and what is pleasurable (distractions and entertainment). In the past, these extremes of boredom assailed mostly those in the upper classes. Now it is something that plagues almost all of us.

There is, however, another possible relationship to boredom and empty time, a fearless one that yields much different results than frustration and escapism. It goes as follows: you have some large goal that you wish to achieve in your life, something you feel that you are destined to create. If you reach that goal, it will bring you far greater satisfaction than the evanescent thrills that come from outside diversions. To get there you will have to learn a craft—educate yourself and develop the proper skills. All human activities involve a process of mastery. You must learn the various steps and procedures involved, proceeding to higher and higher levels of proficiency. This requires discipline and tenacity—the ability to withstand repetitive activity, slowness, and the anxiety that comes with such a challenge.

Once you start down this path, two things will happen: First, having the larger goal will lift your mind out of the moment and help you endure the hard work and drudgery. Second, as you become better at this task or craft, it becomes increasingly pleasurable. You see your improvement; you see connections and possibilities you hadn't noticed before. Your mind becomes absorbed in mastering it further, and in this absorption you forget all your problems—fears for the future or people's nasty games. But unlike the diversion that comes from outside sources, this one comes from within. You are developing a lifelong skill, the kind of mental discipline that will serve as the foundation of your power.

To make this work you must choose a career or a craft that excites you in some deep way. You are creating no dividing line between work and pleasure. Your pleasure comes in mastering the process itself, and in the mental immersion it requires.

In the hood, most of the jobs that are available offer low money and the kind of menial work that leads to no real skills. Even hustling is tedious and not really a path with a future. In the face of this reality, people can go in one of two directions—they can seek to escape this reality through drugs, alcohol, gang activity, or whatever immediate pleasures can be had; or they can get out of the cycle by developing an intense work ethic and discipline. The types who go in the latter direction have a deep hunger for power and a sense of urgency. Nipping at their heels at all times is the possibility of a life of crap jobs or dangerous distractions. They teach themselves to be patient and to practice something. They have learned from early on, through their jobs or through hustling, to endure the long, boring stretches of time that are necessary to master a process. They do not whine or seek to escape this reality, but instead see it as a means to freedom.

For those of us who do not grow up in such an environment, we do not feel this urgent connection between discipline and power. Our jobs are not so dull. Some day they may lead to something really good, or so we think. We have developed some discipline at school or on the job, and it's enough. But we are in fact deluding ourselves. More often than not our jobs are something that we endure; we live for our time off and dream of the future. We are not engaged in the daily activity of the job with our full mental powers because it is not as exciting as life outside work. We develop less and less tolerance for dull moments and repetitive activity. If we happen to lose our job or want something else, we suddenly have to confront the fact that we do not have the requisite patience to make the proper change. Before it is too late we must wake up and realize that real power and success can come only through mastering a process, which in turn depends on a foundation of discipline that we are constantly keeping sharp.

The fearless types in history inevitably display in their lives a higher tolerance than most of us for repetitive, boring tasks. This allows them to excel in their field and master their craft. Part of this comes from seeing early on in life the tangible results that come from such rigorousness and patience. In this vein, the story of Isaac Newton is particularly illuminating. In early 1665 he was a twenty-three-year-old student at Cambridge University, on the verge of taking his exams to be a scholar in mathematics, when suddenly the plague broke out in London. The deaths were horrific and multiplied by the day; many Londoners fled to the countryside where they spread the plague far and wide. By that summer, Cambridge was forced to close, and its students dispersed in all directions for their safety.

For these students, nothing could have been worse. They were forced to live in scattered villages and experienced intense fear and isolation for the next twenty months, as the plague raged throughout England. Their active minds had nothing to seize upon and many went mad with boredom. For Newton, however, the plague months represented something entirely different. He returned to his mother's home in Woolsthorpe, Lincolnshire. At Cambridge he had been bothered by a series of mathematical problems that tortured not only him but his professors as well. He decided he would spend the time in Woolsthorpe working over such problems. He had carried with him a large number of books on mathematics that he had accumulated, and he proceeded to study them in intense detail. He went over the same problems, day after day, filling notebooks with endless calculations.

When the sky was clear he would wander outside and continue these musings, seated in the apple orchards surrounding the house. He would look up at an apple dangling on a branch, the same size to his eye as the moon above, and he would ponder the relationship between the two—what held the one on the tree and the other within the earth's orbit—leading him to ideas about gravity. Staring at the sun and its optical effect on everything around him, he began to conduct his own experiments on the movement and properties of light itself. His mind flowed naturally from problems of geometry to how it all related to motion and mechanics.

The deeper he went into these studies, the more he would see connections and have sudden insights. He solved problem after problem, his enthusiasm and momentum quickening as he realized the powers he was unleashing in himself. While the others were paralyzed with fear and boredom, he passed the entire twenty months without a thought of the plague or any worries for the future. And in that time, he essentially created modern mathematics, mechanics, and optics. It is generally considered the most prolific, concentrated period of scientific thinking in the history of mankind. Of course, Isaac Newton possessed a rare mind, but at Cambridge nobody had suspected him of such mental powers. It took this period of forced isolation and repetitive labor to transform him into a genius.

When we look at those who stand out in history, we tend to focus on their achievements. From such an angle, it is easy for us to be dazzled and see their success as stemming from genetics and perhaps some social factors. They are gifted. We could never reach their level, or so we think. But we are choosing to ignore that telling period in their lives, when each and every one of them underwent a rather tedious apprenticeship in their field. What kept them going was the power they quickly discovered through mastery of certain steps. Sudden insights came to them that seem like genius to us, but are actually part of any intense learning process.

If only we were to study that part of their lives as opposed to the legends they later became, we would understand that we too could have some or all of that power by a patient immersion in any field of study. Many people cannot handle the boredom this might entail; they fear starting out on such an arduous process. They prefer their distractions, dreams, and illusions, never aware of the higher pleasures that are there for those who choose to master themselves and a craft.

||||||||||||||||||||||||||||||||||||||||||||||||||||||||||||||||||||||||||||||||||||||||||||||||||||||||||||||||||||||||||||||||||||

Keys to Fearlessness

||||||||||||||||||||||||||||||||||||||||||||||||||||||||||||||||||||||||||||||||||||||||||||||||||||||||||||||||||||||||||||||||||||

ALL OF MAN'S TROUBLES COME FROM NOT KNOWING HOW TO SIT STILL, ALONE IN A ROOM.

—Blaise Pascal

As children learning language, we all undergo the same process. At first we experience a level of frustration—we have desires and needs we wish to express, but we lack the words. Slowly we pick up phrases and absorb patterns of speech. We accumulate vocabulary, word by word. Some of this is tedious but we are impelled by our intense curiosity and hunger for knowledge. At a certain point we attain a level of fluency in which we can communicate as fast as we think. Soon we don't have to think at all— words come naturally, and at times when we are inspired, they flow out of us in ways we cannot even explain. Learning a language—our own or a foreign one— involves a process that cannot be avoided. There are no shortcuts.

Learning language sets the pattern for all human activities—purely intellectual or physical. To master a musical instrument or a game, we begin at the lowest level of competence. The game seems boring as we have to learn the rules and play on a simple level. As with learning language, we feel frustrated. We see others play well and we imagine how that could feel, but we are locked in this mode of tedious practice and repetition. At such a point we either give in to our frustration and give up the process, or we proceed, intuiting the power that lies just around the corner. Slowly our ability rises and the frustration lowers. We don't need to think so much; we are surprised by our fluency and connections that come to us in a flash.

Once we reach a certain level of mastery, we see there are higher levels and challenges. If we are disciplined and patient, we proceed. At each higher level, new pleasures and insights await us—ones not even suspected when we started out. We can take this as far as we want—in any human activity there is always a higher level to which we can aspire.

For thousands of years this concept of learning was an elemental part of practical wisdom. It was embedded in the concept of mastering a craft. Human survival depended on the construction of instruments, buildings, ships, and more. To build them well, a person had to learn the craft, spending years as an apprentice, advancing step by step. With the advent of the printing press and books that could be distributed widely, this discipline and patience was then applied to education— to formally gaining knowledge. Those who posed as people who possessed learning, without the years of accumulating knowledge, were thought of as charlatans and quacks, to be despised.

Today, however, we have reached a dangerous point in which this elemental wisdom is being forgotten. Much of this is due to the destructive side of technology. We all understand its immense benefits and the power it has brought us. But with the intense speed and ease with which we can get what we want, a new pattern of thinking has evolved. We are by nature creatures of impatience. It has always been hard for us to want something and not have the capacity to get it. The increased speed from technology accentuates this childish aspect of our character. The slow accumulation of knowledge seems unnecessarily boring. Learning should be fun, fast, and easy. On the Internet we can make instant connections, skimming along the surface from one subject to the next. We come to value breadth of knowledge over depth, the power to move here or there rather than digging deeper to the source of a problem and finding out how things tick.

We lose a sense of process. In such an atmosphere, charlatans sprout like weeds. They offer the age-old myth of the quick transformation—the shortcut to power, beauty, and success—in the form of books, CDs, seminars, ancient "secrets" brought back to life. And they find many suckers on which to prey.

This new pattern of thinking and learning is not progress. It creates a phenomenon that we shall call the "short-circuit." To reach the end of anything, to master a process, requires time, focus, and energy. When people are so distracted, their minds constantly moving from one thing to another, it becomes increasingly difficult to maintain concentration on one thing for a few hours, let alone for months and years. Under this influence, the mind will tend to short-circuit; it will not be able to go all the way to the end of a task. It will want to move on to something else that seems more enticing. It becomes hard to make things well when the focus is broken—which is why we find a gradual increase in products that are shoddy, made with less and less attention to detail.

Understand: the real secret, the real formula for power in this world, lies in accepting the ugly reality that learning requires a process, and this in turn demands patience and the ability to endure drudge work. It is not sexy or seductive at first glance, but this truth is based on something real and substantial—an age-old wisdom that will never be overturned. The key is the level of your desire. If you are really after power and mastery, then you will absorb this idea deeply and engrave it in your mind: there are no shortcuts. You will distrust anything that is fast and easy. You will be able to endure the initial months of dull, repetitive labor, because you have an overall goal. This will prevent you from short-circuiting, knowing many things but mastering none of them. In the end, what you really will be doing is mastering yourself—your impatience, your fear of boredom and empty time, your need for constant fun and amusement.

The following are five principal strategies for developing the proper relationship to process.

PROGRESS THROUGH TRIAL AND ERROR

Based on his street fighting as a teenager, Jack Johnson had the feeling that he could some day become a great boxer. But he was black and poor, too poor to afford a trainer. And so in 1896, at the age of eighteen, he began a rather remarkable process. He looked for any conceivable fight he could have in the ring, with any kind of opponent. In the beginning he suffered some terrible beatings from boxers who used him as sparring material. But since this was his only form of education, he quickly learned to become as evasive as possible, to prolong the fights so he could learn.

At the time, fights could go a full twenty rounds, and Johnson's goal was always to drag them out to the maximum. In that time he would carefully study his opponents. He observed how some types would move in familiar patterns and how others would telegraph their punches. He could categorize them by the look in their eye and their body language. He learned to provoke some into a rage so he could study their reactions; others he lulled to sleep with a calm style, to see the effects of this as well.

Johnson's method was quite painful—it meant fifteen to twenty bouts a year. He suffered innumerable hard blows. Even though he could knock out most of his opponents, he preferred to be evasive and learn on the job. This meant hearing endless taunts from the mostly white audiences that he was a coward. Slowly, however, it began to pay off. He faced such a variety of foes that he became adept at recognizing their particular style the instant the fight began. He could sense their weaknesses and when exactly he should move in for the kill. He accustomed himself—mentally and physically—to the pace of a long, grueling bout. He gained an intuitive feel for the space of the boxing ring itself, and how to maneuver and exhaust his opponents over the course of twenty rounds. Many of them later confessed that he seemed to have the ability to read their minds; he was always a step ahead. Following this path, within a few short years Johnson transformed himself into the heavyweight champion of the world and the greatest fighter of his era.

Too often our concept of learning is to absorb ideas from books, to do what others tell us to, and perhaps to do some controlled exercises. But this is an incomplete and fearful concept of learning—cut off from practical experience. We are creatures who make things; we don't simply imagine them. To master any process you must learn through trial and error. You experiment, you take some hard blows, and you see what works and doesn't work in real time. You expose yourself and your work to public scrutiny. Your failures are embedded in your nervous system; you do not want to repeat them. Your successes are tied to immediate experience and teach you more. You come to respect the process in a deep way because you see and feel the progress you can make through practice and steady labor. Taken far enough, you gain a fingertip feel for what needs to be done because your knowledge is tied to something physical and visceral. And having such intuition is the ultimate point of mastery.

MASTER SOMETHING SIMPLE

Often we have a general feeling of insecurity because we have never really mastered anything in life. Unconsciously we feel weak and never quite up to the task. Before we begin something, we sense we will fail. The best way to overcome this once and for all is to attack this weakness head-on and build for ourselves a pattern of confidence. And this must be done by first tackling something simple and basic, giving us a taste for the power we can have.

Demosthenes—one of the greatest political figures in ancient Athens—followed such a path, determined to overcome his intense fear of public speaking. As a child, he was frail and nervous. He talked with a stammer and always seemed out of breath. He was constantly ridiculed. His father had died when he was young, leaving him a nice sum of money, but his guardians quickly stole all of it. He decided to become a lawyer and eventually take the perpetrators to court on his own. But a lawyer needed to be an eloquent speaker, and he was an abysmal failure at that. He decided he would give up law—it seemed too difficult. With what little money he had, he would retire from the world and attempt somehow to master his speech impediment. At least then he could take on some kind of public career.

He built an underground study where he could practice alone. He shaved half his head so he would be too embarrassed to go out in public. To overcome his stammer, he walked along the beach with his mouth full of pebbles, forcing himself to speak without stopping, louder and more forcefully than the waves. He wrote speeches that he then recited while running up steep slopes, to develop better breathing techniques. He installed a looking glass in his study, allowing him to monitor closely the looks on his face as he declaimed. He would engage in conversations with visitors to his house and gauge how each word or intonation would affect them. Within a year of such dedicated practice, he had completely eliminated his stammer and had transformed himself into a more than adequate orator. He decided to return to law after all. With each new case that he won, his confidence rose to new heights.

Understanding the value of practice, he then worked on improving the delivery of his speeches. Slowly he transformed himself into the supreme orator of ancient Athens. This new-found confidence translated into everything he did. He became a leading political figure, renowned for his fearlessness in the face of any foe.

When you take the time to master a simple process and overcome a basic insecurity, you develop certain skills that can be applied to anything. You see instantly the reward that comes from patience, practice, and discipline. You have the sense that you can tackle almost any problem in the same way. You create for yourself a pattern of confidence that will continue to rise.

INTERNALIZE THE RULES OF THE GAME

As a law student at Howard University in the early 1930s, Thurgood Marshall could contemplate many injustices that blacks experienced in the United States, but the one that burned in him most deeply was the vast inequalities in education. He had toured the South on fact-finding missions for the NAACP and had seen firsthand the abysmal quality of schools set apart for blacks. And he had felt this injustice himself. He had wanted to go to the University of Maryland, near his home—it had an excellent law school. But black students were not admitted there, no matter their academic record. They were directed towards black universities such as Howard, which at the time were inferior. Marshall vowed that some day, in some way, he would help take this unjust system apart.

Upon graduation from Howard in 1933, he faced a crucial decision for his future. He had been offered a scholarship at Harvard University to study for an advanced law degree. This represented an incredible opportunity. He could carve out for himself a nice position within the academic world and promote his ideas in various journals. It was also the middle of the Depression, and jobs for black people were few and far between; a degree from Harvard would ensure him a prosperous future. But something impelled Marshall in the opposite direction; he decided instead to set up a private practice in Baltimore and learn from the ground up how the justice system worked. At first it seemed a foolish decision—he had little work and his debts were mounting. The few cases he had, he lost and he could not figure out why. The justice system seemed to have its own rules and codes to which he had no access.

Marshall decided to employ a unique strategy to overcome this. First, he made sure that his legal briefs were masterpieces of research and detail, without any errors or erasures. He made a point of always dressing in the most professional manner and acting with the utmost courtesy, without appearing to bow and scrape. In other words, he gave no one the slightest pretext for judging against him. In this way, he defused suspicion, began to win a few cases, and gained entrée to the world of white lawyers. Now he studied that world closely. He saw the importance of certain connections and friendships, power networks he had not known about before. He recognized that certain judges required certain treatment. He learned to talk the language and fit in socially as best he could. He found out that in most cases, it was best to argue on points of narrow procedure rather than on grand concepts.

Knowing how to maneuver within these rules and conventions, he began to win more and more cases. In 1935 he took on the University of Maryland on behalf of a black student who had been denied admission to its law school, and won. From then on, he used his knowledge to take on all forms of segregation in the education system, culminating in 1954 with his greatest triumph of all, arguing before the U.S. Supreme Court the case known as *Brown v. Board of Education.* The court's decision in his favor effectively ended any basis for educational segregation in the United States. What Marshall (who would later become the first African American appointed to the Supreme Court) had learned by immersing himself in the white-controlled justice system of his time is that the social process is just as important as the legal or technical one. This was not something taught in law school and yet learning it was the key to his ability to function within the system and advance the cause for which he was fighting.

Understand: when you enter a group as part of a job or a career, there are all kinds of rules that govern behavior—

values of good and bad, power networks that must be respected, patterns to be followed for successful action. If you do not patiently observe and learn them well, you will make all kinds of mistakes without knowing why or how. Think of social and political skills as a craft that you must master as well as any other. In the initial phase of your apprenticeship you must do as Marshall did and mute your colors. Your goal here is not to impress people with your brilliance but to learn these conventions from the inside. Watch for telling mistakes that others have made in the group and for which they have paid a price—that will reveal particular taboos within the culture. With a deepening knowledge of these rules you can begin to maneuver them for your purpose. If you find yourself confronting an unjust and corrupt system, it is much more effective to learn its codes from the inside and discover its vulnerabilities. Knowing how it works, you can take it apart—for good.

ATTUNE YOURSELF TO THE DETAILS

As a young student-artist in late fifteenth-century Italy, Michelangelo had to confront a personal limitation. He had grand concepts of things he wanted to paint and sculpt, but not the requisite skill. He looked at the masterpieces of other artists and wanted his own work to have a similar aura and effect, but he was frustrated at the flatness and conventionality of what he created. He tried an experiment: he began to copy his favorite masterworks down to the smallest brush stroke, and he discovered that the effect he had so admired was embedded in certain details—the way these artists were able to make figures or landscapes come to life by their intense attention to the fine points. And so began a remarkable apprenticeship to his craft that would last the rest of his life and completely alter his way of thinking.

In creating his sculptures, he became obsessed with bone structure, but the books and techniques on the subject seemed woefully inadequate. He started dissecting human corpses, one after another. This gave him a profound feel for human anatomy that he could now reproduce in his work. He developed an interest in texture, how each kind of fabric would fold in its own way. He worked on perfecting his reproduction of clothes. He extended these studies of detail to animals and how they moved. When he was commissioned to do his larger pieces, he avoided that old temptation of beginning with some grand concept—instead he looked at the material he was to work with, the space, the individual figures that might comprise it, and from there he would conceive the overall shape and effect. In this intense attention to detail, Michelangelo seemed to have discovered the secret for making his figures come to life in a way that exceeded any other artist of his time.

Often when you begin a project of any kind, it is from the wrong end. You tend to think first of what you want to accomplish, imagining the glory and money it will bring you if it succeeds. You then proceed to make this concept come to life. But as you go forward you often lose patience, because the small steps to get there are not nearly as exciting as the ambitious visions in your head. You must try instead the opposite approach, which can lead to very different results. You have a project you wish to bring to life, but you begin by immersing yourself in the details of the subject or field. You look at the materials you have to work with, the tastes of your target audience, and the latest technical advances in the field. You take pleasure in going deeper and deeper into these fine points—your research is intense. From this knowledge, you shape the project itself, grounding it in reality rather than in airy concepts in your head. Operating this way helps you slow your mind down and develop patience for detailed work, an essential skill for mastering any craft.

REDISCOVER YOUR NATURAL PERSISTENCE

This is the dilemma we all face: to accomplish anything worthwhile in life generally takes some time— -often in blocks of years. But we are creatures who find it very hard to manage such long periods. We are immersed in the day-to-day; our emotions fluctuate with each encounter. We have immediate desires we are constantly working to satisfy. In that long period of time that we need to reach a goal, we are assailed by a thousand distractions and temptations that seem more interesting. We lose sight of our objectives and end up following some detour. This is the source of so many of the failures in our lives.

To force yourself past any obstacle or temptation, you must be persistent. As children we all had this quality because we were single-minded; you must simply rediscover and redevelop this character trait. First, you must understand the role that your energy level plays in mastering a process and bringing something to completion. If you take on added goals or new tasks, your focus will be broken up and you will never attain what you wanted in the first place. You cannot persist on two or three paths, so avoid that temptation. Second, try breaking things up into smaller blocks of time. You have a large goal, but there are steps along the way, and steps within the steps. These steps represent months instead of years. Reaching these smaller goals gives you a sense of tangible reward and progress. This will make it easier for you to resist any diversions along the way and fearlessly push ahead. Remember: anything will give way to a sustained, persistent attack on your part.

Reversal of Perspective

We generally experience boredom as something painful and to be avoided at all costs. From childhood on, we develop the habit of immediately looking for some activity to kill the feeling. But this activity, if repeated often enough, becomes boring as well. And so for our entire lives we must search and search for novel amusements—new friends, new trends to latch on to, new forms of entertainment, new religions or causes to believe in. This search might lead us to change our careers and set us on a path of meandering here and there, in search of something to dull the sensation. But in all of these cases, the root of the problem is not boredom itself but our relationship to it.

Try to look at boredom from the opposite perspective—as a call for you to slow yourself down, to stop searching for endless distractions. This might mean forcing yourself to spend time alone, overcoming that childish inability to sit still. When you work through such self-imposed boredom, you will find your mind clicks into gear—new and unexpected thoughts will come to you to fill the void. To feel inspired you must first experience a moment of emptiness. Use such moments to assess the day that went by, to measure where you are headed. It is a relief to not feel that constant need for outside entertainment.

On a higher level of this reeducation, you might choose a book to overcome your boredom, but instead of reading being a passive process of diversion, you actively mentally engage the author in an argument or discussion, making the book come to life in your head. At a further point, you take up a side activity—cultural or physical—that requires a repetitive process to master. You discover a calming effect in the repetitive element itself. In this way, boredom becomes your great ally. It helps you to slow things down, develop patience and self-discipline. Through this process you will be able to withstand the inevitable empty moments of life and convert them into your own private pleasures.

NOW THERE ARE . . . INDIVIDUALS WHO WOULD RATHER PERISH THAN WORK WITHOUT TAKING *PLEASURE* IN THEIR WORK; THEY ARE CHOOSY . . . AND HAVE NO USE FOR AMPLE REWARDS IF THE WORK IS NOT ITSELF THE REWARD OF REWARDS. . . . THEY DO NOT FEAR BOREDOM AS MUCH AS WORK WITHOUT PLEASURE; INDEED, THEY NEED A LOT OF BOREDOM IF *THEIR* WORK IS TO SUCCEED. FOR . . . ALL INVENTIVE SPIRITS, BOREDOM IS THAT DISAGREEABLE "LULL" OF THE SOUL THAT PRECEDES A HAPPY VOYAGE AND CHEERFUL WINDS.

—Friedrich Nietzsche

CHAPTER 9

Push Beyond Your Limits—Self-Belief

YOUR SENSE OF WHO YOU ARE WILL DETERMINE YOUR ACTIONS AND WHAT YOU END UP GETTING IN LIFE. IF YOU SEE YOUR REACH AS LIMITED, THAT YOU ARE MOSTLY HELPLESS IN THE FACE OF SO MANY DIFFICULTIES, THAT IT IS BEST TO KEEP YOUR AMBITIONS LOW, THEN YOU WILL RECEIVE THE LITTLE THAT YOU EXPECT. KNOWING THIS DYNAMIC, YOU MUST TRAIN YOURSELF FOR THE OPPOSITE—ASK FOR MORE, AIM HIGH, AND BELIEVE THAT YOU ARE DESTINED FOR SOMETHING GREAT. YOUR SENSE OF SELF-WORTH COMES FROM YOU ALONE—NEVER THE OPINION OF OTHERS. WITH A RISING CONFIDENCE IN YOUR ABILITIES, YOU WILL TAKE RISKS THAT WILL INCREASE YOUR CHANCES OF SUCCESS. PEOPLE FOLLOW THOSE WHO KNOW WHERE THEY ARE GOING, SO CULTIVATE AN AIR OF CERTAINTY AND BOLDNESS.

The Hustler's Ambition

LET ME POINT OUT TO YOU THAT FREEDOM IS NOT SOMETHING THAT ANYBODY CAN BE GIVEN; FREEDOM IS SOMETHING PEOPLE TAKE AND PEOPLE ARE AS FREE AS THEY WANT TO BE.

—James Baldwin

Curtis Jackson's mother, Sabrina, had one powerful ambition in her life—to somehow make enough money to move her and her son far away from the hood. She had had Curtis when she was fifteen, and the only reasonable outlet at that age for making any good money was dealing drugs. It was a particularly dangerous life for a female hustler, and so she built up an intimidating presence to protect herself. She was tougher and more fearless than many of the male dealers. Her only soft spot was her son—she wanted a different fate for him than hustling. To shelter him from the life she led, she had him stay with her parents in Southside Queens. As often as possible, she would show up with presents for the boy and to keep an eye on him. Some day soon they would move to a better place.

As part of a drug beef, Sabrina was murdered at the age of twenty-three, and from that moment on, it looked like Curtis's fate in this world had been sealed. He was now essentially alone—no parents or real mentor to give him a sense of direction. It seemed almost certain that the following scenario would play itself out: He would drift towards life on the streets. To prove his toughness, he would eventually have to resort to violence and crime. He would find his way into the prison system, and he would probably return for several stints. His life would basically be confined to this neighborhood, and as he got older he could turn to drugs or alcohol to see him through, or at best a series of menial jobs. All the statistics on parentless children growing up in such an environment pointed towards this limited and bleak future.

And yet in his mind something much different was taking shape. With his mother gone, he spent more and more time alone and began to indulge himself in fantasies that carried him far beyond his neighborhood. He saw himself as a leader of some sort, perhaps in business or in war. He visualized in great detail the places where he would live, the cars he would drive, the outside world he would some day explore. It was a life of freedom and possibility. But these were not mere fantasies—they were real; they were destined to happen. He could see them clearly. Most important, he felt that his mother was looking after him— her energy and ambition were inside him now.

Oddly enough, he would follow in her footsteps with the same plan—to hustle and get out of the game. To avoid her fate he forged an intense belief that nothing could stop him—not a gunshot, the schemes of other hustlers, or the police. These streets would not confine him.

In May of 2000, Curtis (now known as 50 Cent) somehow survived the nine bullets that a hired assassin had pumped into his body. The timing of the attack had been particularly poignant—after years of hustling on the streets and in music, his first album had been about to come out. But then in the aftermath of the shooting, Columbia Records canceled the album and dropped him from the label. He would have to start all over. In the months to come, as he lay in bed recovering from his wounds, he began to reconstruct himself mentally, much as he had done after the murder of his mother. He saw in his mind, in even fresher detail than ever, the path he would now have to take. He would conquer the rap world with a mix-tape campaign the likes of which no one had ever seen before. It would come from his intense energy, his persistence, the even tougher sound he would create, and the image he would now project of an indestructible gangsta.

Within a year of the shooting, he was on his way to making this vision a reality. His first songs hit the streets and created a sensation. As he progressed on this path, however, he saw one very large impediment still blocking his path: the assassins were looking to finish the job and they could show up at any moment. Fifty was forced to keep a low profile and stay on the run, but this feeling of being hunted was intolerable. He would not live this way, and so he decided that what he needed was a group of tight-knit disciples who would help protect him and overcome his sense of isolation.

To make this a reality, he told his closest friends to convene a meeting in his grandparents' house in Southside Queens. They were to invite his most fervent fans in the neighborhood— the young men whom they knew to be loyal and dependable. And they should all bring guns to help secure the street before Fifty showed up.

When Fifty finally entered the living room of his grandparents' house the day of the meeting, he could feel the energy and excitement. The space was filled with over twenty young men, all ready to do his bidding. He began by painting for them his precise vision of the future. His music now was hot, but it was going to get a lot hotter. Within two more years, he was certain to land a major record deal. In his head, he already could hear the songs for his first record, visualize the cover and the overall concept—it was to be the story of his life. This record, he assured them, would be an astronomical success, because he had figured out a kind of formula for how to make and market hit songs. He was not the usual rap star, he explained. He was not in this for the bling or the attention, but for the power. He would take the money from the record sales to establish his own businesses. This was destiny—everything in his life was meant to happen as it did, including the assassination attempt, including this very meeting that afternoon.

He was going to forge a business empire and he wanted to

take all of them with him. Whatever any of them wanted, he would provide, as long as they proved themselves dependable and shared his sense of purpose. They could be rappers on the record label he would establish or road managers for his tours; or they could go to college and get a degree—he would pay for it all. You are like my pack of wolves, he explained, but none of this will happen if the alpha wolf is killed. What he asked for was their help—in providing security, in keeping him in touch with what was happening on the streets, and in doing some of the legwork for the promotion and distribution of his mix-tapes. He needed followers and he had chosen them.

Almost all of them agreed to the proposal, and over the years to come many of them stayed on to gain important positions within his expanding empire. And if they ever stopped to think about it, it was uncanny how close the future had come to resemble the picture he had painted so many years before.

By 2007, after the tremendous success of his first two records, Fifty began to sense a problem looming on the horizon. He had created an image for the public, a Fifty myth that centered on his tough and menacing presence and his indestructibility. This was projected in his videos and interviews, and the photos of him with his glare and tattoos. Most of it was real, but it was all heightened for dramatic effect. This image had brought him a great deal of attention, but it was turning into an elaborate trap. To prove to his fans that he was still the same Fifty, he would have to keep upping the ante, engaging in more and more outrageous antics. He could not afford to seem like he was going soft. But it was not real *to him* anymore. He had moved on to a different life, and to stay rooted in this past image would prove to be the ultimate limit to his freedom. He would be trapped in the past and the prisoner of the very image he had created. It would all grow stale and his popularity would wane.

In each phase of his life he had found himself challenged by some seemingly insurmountable obstacle—surviving on the streets without parents to guide him, keeping away from the violence and time in prison, eluding the assassins on his heels, etc. If at any moment he had doubted himself or accepted the normal limits to his mobility, he would be dead or powerless, which was as good as dead in his mind. What had saved him in each case was the intensity of his ambition and self-belief.

Now was not the time to get complacent or have doubts about the future. He would have to transform himself again. He would have his signature tattoos removed; perhaps he would also change his name again. He would create a new image and myth to fit this period of his life—part business mogul, part power broker, slowly withdrawing from the public eye and flexing his muscles behind the scenes. This would surprise the public, keep him a step ahead of their expectations, and remove yet another barrier to his freedom. Reinventing himself in this way would be the ultimate reversal of the fate that seemed to await him after the death of his mother.

||||||||||||||||||||||||||||||||||||||||||||||||||||||||||||||||||||||||||||||||||||||||||||||||||||||||||||||||||||||||||||||||||

The Fearless Approach

||||||||||||||||||||||||||||||||||||||||||||||||||||||||||||||||||||||||||||||||||||||||||||||||||||||||||||||||||||||||||||||||||

YOUR OPINION OF YOURSELF BECOMES YOUR RE-
ALITY. IF YOU HAVE ALL THESE DOUBTS, THEN NO
ONE WILL BELIEVE IN YOU AND EVERYTHING WILL GO
WRONG. IF YOU THINK THE OPPOSITE, THE OPPOSITE
WILL HAPPEN. IT'S THAT SIMPLE.

—50 Cent

When you were born, you entered this world with no identity or ego. You were simply a bundle of chaotic impulses and
desires. But slowly you acquired a personality that you have
more or less built upon over the years. You are outgoing or shy,
bold or skittish—a mix of various traits that defines you. You
tend to accept this personality as something very real and established. But much of this identity is shaped and constructed
by outside forces—the opinions and judgments of hundreds of
other people who have crossed your path over the years.

This process began with your parents. As a child you paid
extra close attention to what they said about you, modeling
your behavior to win their approval and love. You closely mon-
itored their body language to see what they liked and didn't
like. Much of this had a tremendous impact on your evolution.
If, for example, they commented about your shyness, it could
easily strengthen any tendencies you had in that direction. You
suddenly became aware of your own awkwardness and it stuck

inside you. If they had said something different, trying to encourage you in your social skills and draw you out, it might have had a much different impact. Either way, shyness is a fluid quality—it fluctuates according to the situation and the people you are around. It should never be felt as a set personality trait. And yet these judgments from parents, friends, and teachers are given inordinate weight and become internalized.

Many of these criticisms and opinions are not objective at all. People want to see certain qualities in you. They project onto you their own fears and fantasies. They want you to fit a conventional pattern; it is frustrating and often frightening for people to think they cannot figure someone out. Behavior that is considered abnormal or different, which may very well be coming from somewhere deep within you, is actively discouraged. Envy plays a role as well—if you are too good at something, you might be made to feel strange or undesirable. Even the praise of others is often designed to hem you in to certain ideals they want to see in you. All of this shapes your personality, limits your range of behavior, and becomes like a mask that hardens on your face.

Understand: you are in fact a mystery to yourself. You began life as someone completely unique—a mix of qualities that will never be repeated in the history of the universe. In your earliest years, you were a mass of conflicting emotions and desires. Then something foreign to you is placed over this reality. Who you are is much more chaotic and fluid than this surface character; you are full of untapped potential and possibility.

As a child you had no real power to resist this process, but as an adult you could easily rebel and rediscover your individuality. You could stop deriving your sense of identity and self-worth from others. You could experiment and push past the limits people have set for you. You could take action that is different from what they expect. But that is to incur a risk. You are being unconventional, perhaps a bit strange in the eyes of those who know you. You could fail in this action and be ridiculed. Conforming to people's expectations is safer and more comfortable, even if doing so makes you feel miserable and confined. In essence, you are afraid of yourself and what you could become.

There is another, fearless way of approaching your life. It begins by untying yourself from the opinions of others. This is not as easy as it sounds. You are breaking a lifelong habit of continually referring to other people when measuring your value. You must experiment and feel the sensation of not concerning yourself with what others think or expect of you. You do not advance or retreat with their opinions in mind. You drown out their voices that often translate into doubts inside you. Instead of focusing on the limits you have internalized, you think of the potential you have for new and different behavior. Your personality can be altered and shaped by your conscious decision to do so.

We barely understand the role that willpower plays in our actions. When you raise your opinion of yourself and what you are capable of it has a decided influence on what you do. For instance, you feel more comfortable taking some risk, knowing that you are always able to get back up on your feet if it fails.

Taking this risk will then make your energy levels rise—you have to meet the challenge or go under, and you will find untapped reservoirs of creativity within you. People are drawn to those who act boldly, and their attention and faith in you will have the effect of heightening your confidence. Feeling less confined by doubts, you give freer rein to your individuality, which makes everything you do more effective. This movement towards confidence has a self-fulfilling quality that is impossible to deny.

Moving towards such self-belief does not mean you cut yourself off from others and their opinions of your actions. You must take constant measure of how people receive your work, and use to maximum effect their feedback (see chapter 7). But this process must begin from a position of inner strength. If you are dependent on their judgments for your sense of worth, then your ego will always be weak and fragile. You will have no center or sense of balance. You will wilt under criticisms and soar too high with any praise. Their opinions are merely helping you shape your work, not your self-image. If you make mistakes, if the public judges you negatively, you have an unshakable inner core that can accept such judgments, but you remain convinced of your own worth.

In impoverished environments like the hood, people's sense of who they are and what they deserve is continually under attack. People from the outside tend to judge them for where they come from—as violent, dangerous, or untrustworthy—as if the accident of where they were born determines who they are. They tend to internalize many of these judgments and perhaps deep inside feel that they don't deserve much of what is considered good in this world. Those from the hood who want to overcome this pronouncement of the outside world have to fight with double the energy and desperation. They have to convince themselves first that they are worth much more and can rise as far as they want, through willpower. The intensity of their ambition becomes the deciding factor. It has to be supremely high. That is why the most ambitious and confident figures in history often emerge from the most impoverished and arduous of circumstances.

For those of us who live outside such an environment, "ambition" has almost become a dirty word. It is associated with such historical types as Richard III or Richard Nixon. It reeks of insecurity and evil deeds to reach the top. People who want power so badly must have psychological problems, or so we think. Much of this social prudery around the idea of power and ambition comes from an unconscious guilt and desire to keep other people down. To those occupying a position of privilege, the ambitiousness of those from below seems like something scary and threatening.

If you come from relative prosperity, you are more than likely tainted with some of this prejudice and you must extirpate it as much as possible. If you believe ambition is ugly and needs to be disguised or repressed, you will have to tiptoe around others, making a show of false humility, in two minds every time you contemplate some necessary power move. If you see it as beautiful, as the driving force behind all great human accomplishments, then you will feel no guilt in raising your level of ambition as high as you want and pushing aside those who block your path.

One of the most fearless men in history has to be the great nineteenth-century abolitionist Frederick Douglass. He was born into the cruelest system known to man—slavery. It was designed in every detail to destroy a person's spirit. It did so by separating people from their families, so they could develop no real emotional attachments in their lives. It used constant threats and fear to break any sense of free will, and it made sure that slaves were kept illiterate and ignorant. They were to form only the lowest opinions of themselves. Douglass himself suffered all of these fates as a child, but somehow from his earliest years he believed that he was worth much more, that something powerful had been crushed but that it could spring back to life. As a child he saw himself escaping the clutches of slavery some day, and he nourished himself on that dream.

Then in 1828, at the age of ten, Douglass was sent by his master to work in the home of a son-in-law in Baltimore, Maryland. Douglass read this as some kind of providence working in his favor. It meant he would escape the hard labor on the plantation and have more time to think. In Baltimore, the mistress of the house was constantly reading the Bible, and one day he asked her if she would teach him to read. She happily obliged and he quickly learned. The master of the house heard of this and severely upbraided his wife—a slave must never be allowed to read and write. He forbade her to continue with the teachings. Douglass, however, could now manage on his own, getting books and dictionaries for himself on the sly. He memorized famous speeches, which he could go over in his mind at any time of day. He saw himself becoming a great orator, railing against the evils of slavery.

With growing knowledge of the outside world, he came to resent even more bitterly the life he was forced to lead. This infected his attitude, and his owners sensed it. At the age of fifteen he was sent to a farm run by a Mr. Covey, whose sole task in life was to break the spirit of a rebellious slave.

Covey, however, was not successful. Douglass had already created in his mind an identity for himself that would not match what Covey wanted to impose on him. This image of his own high value, believed in with all his energy, would become reality. He maintained his inner freedom and his sanity. He converted all of the whippings and mistreatment into a spur for him to escape to the North; it gave him more material to some day share with the world on the evils of slavery. Several years later, Douglass managed to escape to the north. There he became a leading abolitionist, eventually founding his own newspaper and always pushing against the limits people tried to impose on him.

Understand: people will constantly attack you in life. One of their main weapons will be to instill in you doubts about yourself—your worth, your abilities, your potential. They will often disguise this as their objective opinion, but invariably it has a political purpose—they want to keep you down. You are continually prone to believe these opinions, particularly if your self-image is fragile. In every moment of life you can defy and deny people this power. You do so by maintaining a sense of purpose, a high destiny you are fulfilling. From such a position, people's attacks do not harm you; they only make you angry and more determined. The higher you raise this self-image, the fewer judgments and manipulations you will tolerate. This will translate into fewer obstacles in your path. If someone like Douglass could forge this attitude amid the most unfree of circumstances, then we should surely be able to find our own way to such inner strength.

Keys to Fearlessness

ONE'S OWN FREE, UNTRAMMELED DESIRES, ONE'S OWN WHIM . . . ALL OF THIS IS PRECISELY THAT WHICH FITS NO CLASSIFICATION, AND WHICH IS CONSTANTLY KNOCKING ALL SYSTEMS AND THEORIES TO HELL. AND WHERE DID OUR SAGES GET THE IDEA THAT MAN MUST HAVE NORMAL, VIRTUOUS DESIRES? WHAT MAN NEEDS IS ONLY HIS OWN INDEPENDENT WISHING, WHATEVER THAT INDEPENDENCE MAY COST AND WHEREVER IT MAY LEAD.

—Fyodor Dostoyevsky

In today's world our idea of freedom largely revolves around the ability to satisfy certain needs and desires. We feel free if we can gain the kind of employment we desire, buy the things we want, and engage in a wide range of behavior, as long as it does not harm others. According to this concept, freedom is something essentially passive—it is given and guaranteed to us by our government (often by not meddling in our affairs) and various social groups.

There is, however, a completely different concept of liberty. It is not something that people grant us as a privilege or right. It is a state of mind that we must work to attain and hold on to—with much effort. It is something active and not passive. It comes from exercising free will. In our day-to-day affairs much of our actions are not free and independent. We tend to conform to society's norms in behavior and thinking. We generally act out of habit, without much thought as to why we do things. When we act with freedom, we ignore any pressures to conform; we step beyond our usual routines. Asserting our will and our individuality, we move on our own.

Let us say we have a career that affords us enough money to live comfortably and offers us a reasonable future. But this job is not deeply satisfying; it doesn't lead us anywhere we want to go. Perhaps we also have to deal with a boss who is difficult and imperious. Our fears for the future, our comfortable habits, and our sense of propriety will compel us to stay on. All of these factors are forces that limit and confine us. But at any moment we could let go of the fear and leave the job, not really certain where we are headed but confident we can do better. In that moment we have exercised free will. It initiates from our own deepest desire and need. Once we leave, our mind must rise to the challenge. To continue on this path, we have to take more independent actions, because we cannot depend on habits or friends to see us through. Free action has a momentum of its own.

Many will argue that this idea of active freedom is basically an illusion. We are products of our environment, so they say. If people become successful, it is because they benefited from certain favorable social circumstances—they were in the right place at the right time; they got the proper education and mentoring. Their willpower played a part, no doubt, but a small part. If circumstances were different, so the argument goes, these types would not have had the success they had, no matter how strong their willpower.

All kinds of statistics and studies can be trotted out to support this argument, but in the end this concept is merely a product of our times and the emphasis on passive freedom. It chooses to focus on circumstance and environment, as if the exceptionally free actions of a Frederick Douglass could also be explained by his physiology or the luck he had in learning to read. In the end, such a philosophy *wants* to deny the essential freedom we all possess to make a decision independent of outside forces. It wants to diminish individuality—we are just products of a social process, they imply.

Understand: at any moment you could kick this philosophy and its ideas into the trashcan by doing something irrational and unexpected, contrary to what you have done in the past, an act not possibly explained by your upbringing or nervous system. What prevents you from taking such action is not mommy, daddy, or society, but your own fears. You are essentially free to move beyond any limits others have set for you, to re-create yourself as thoroughly as you wish.

If you had some terribly painful experience in the past, you could choose to let that pain sit there and you could soak in it. On the other hand, you could decide to convert it into anger, a cause to promote, or some form of action. Or you could choose to simply drop it and move on, relishing the freedom and power that that brings you. No one can take away these options or force your response. It is all up to you.

Moving to this more active form of freedom does not mean that you are giving yourself over to wild and ill-considered action. Your decision to alter a career path, for instance, is based on careful consideration of your strengths and deepest desires and the future you want. It comes from thinking for yourself and not accepting what others think about you. The risks you take are not emotional and for the sake of a thrill; they are calculated. The need to conform and please others will always play a role in our actions, consciously or unconsciously. To be completely free is impossible and undesirable. You are merely exploring a freer range of action in your life and the power it could bring you.

What block us from moving in this direction are the pressures we feel to conform; our rigid, habitual patterns of thinking; and our self-doubts and fears. The following are five strategies to help you push past these limits.

DEFY ALL CATEGORIES

As a young girl growing up in Kansas at the turn of the twentieth century, Amelia Earhart felt oddly out of place. She liked to do things her own way—playing rough games with the boys, spending hours by herself reading books, or disappearing on long hikes. She was prone to behavior that others considered strange and unorthodox—at boarding school she was kicked out for walking on the roof in her nightgown. As she got older she felt intense pressure to settle down and be more like other girls. Earhart had an abhorrence, however, of marriage and the constrictions it represented for women, so she looked for a career, trying her hand at all kinds of jobs. But she craved adventure and challenges, and the jobs available to her were menial and mindless.

Then one day in 1920 she went for a short ride in an airplane, and suddenly she knew she had found her calling. She took lessons and became a pilot. In the air she felt the freedom she had always been looking for. Piloting a plane was a constant challenge—physical and mental. She could express the daring side of her character, her love of adventure, as well as her interest in the mechanics of flying.

Female pilots at the time were not taken seriously. The men were the ones who set records and blazed new paths. To combat this, Earhart had to push the limits as far as she could, doing feats of flying that would make headlines and contribute something to the profession. In 1932 she became the first woman to pilot a plane solo across the Atlantic, in what would turn out to be her most death-defying and physically arduous flight. In 1935 she contemplated doing a crossing of the Gulf of Mexico. One of the most famous male pilots of the time told her it was too dangerous and not worth the risk. Feeling there was a challenge in this, she decided to attempt the flight anyway and managed it with relative ease, showing others how it could be done.

If at any moment in her life she had succumbed to the pressure to be more like others, she would have lost that magic that now seemed to follow her when she went her own direction. She decided to continue being herself, whatever the consequences might be. She dressed in her unconventional manner and spoke her mind on political matters, even though that was considered unbecoming. When the famous publicist and promoter George Putnam asked for her hand in marriage, Earhart accepted under the condition that he sign a contract guaranteeing he would respect her desires for maximum freedom within the relationship.

People who met her invariably commented that she was not really masculine or feminine or even androgynous, but completely herself, a unique mix of qualities. It was this part of her that fascinated people and kept her in the limelight. In 1937, she attempted the riskiest flight of her career—to circle the world via the equator, including a stopover on a tiny island in the Pacific. She disappeared somewhere near the island, never to be found, all of which only added to the legend of Earhart as the consummate risk taker who did everything her own way.

Understand: the day you were born you became engaged in a struggle that continues to this day and will determine your success or failure in life. You are an individual, with ideas and skills that make you unique. But people are constantly trying to fit you into narrow categories that make you more predictable and easier to manage. They want to see you as shy or outgoing, sensitive or tough. If you succumb to this pressure, then you may gain some social acceptance, but you will lose the unconventional parts of your character that are the source of your uniqueness and power. You must resist this process at all costs, seeing people's neat and tidy judgments as a form of confinement. Your task is to retain or rediscover those aspects of your character that defy categorization, and to give them even greater play. Remaining unique, you will create something unique and inspire the kind of respect you would never receive from tepid conformity.

CONSTANTLY REINVENT YOURSELF

As a child, the future president John F. Kennedy was extremely frail and prone to illness. He spent much time in various hospitals, and grew up to be rather frail and weak looking. From these experiences he developed a horror of anything that made him feel that he had no control over his life. And one form of powerlessness particularly irked him—the judgments people made of him based on his appearance. They would see him as weak and fragile, underestimating his underlying strength of character. So he initiated a lifelong process of wresting this control from others, constantly re-creating himself and casting the image that he wanted people to see of him.

As a youth, he was perceived as the pleasure-loving son of a powerful father, so at the outbreak of World War II, despite his physical limitations, he enlisted in the navy, determined to show another side of himself. As a lieutenant on a patrol torpedo in the Pacific, his boat was rammed and cut in two by a Japanese destroyer. He proceeded to lead his men to safety in a way that earned him numerous medals for bravery. During this incident he displayed an almost callous disregard for his own life, perhaps in an attempt once and for all to prove his masculinity. In 1946, he decided to run for Congress, and he used his war record to craft the image of a young man who would be an equally fearless fighter for his constituency.

A few years later, as a senator, he realized that many in the public perceived him as a bit of a lightweight—young and unproven. And so yet again, he chose to reinvent himself, this time by writing a book (authored with his speechwriter Theodore Sorenson) called *Profiles in Courage*, cataloguing stories of famous senators who defied convention and achieved great things. The book won a Pulitzer Prize, and more important, it completely altered the image the public had of Kennedy. He was now seen as thoughtful and independent, somehow following the path of the senators he had written about—clearly an intended effect.

In 1960, when Kennedy was running for president, people once again underestimated him. They saw him as the young Catholic liberal senator who could not possibly appeal to the majority of Americans. This time he decided to recast himself as the inspiring prophet who would lead the country out of the doldrums of the Eisenhower era, returning America to its frontier roots and creating a sense of unified purpose. It was an image of vigor and youth (contrary to his still physical weakness) and it proved compelling enough to captivate the public and win the election.

Understand: people judge you by appearances, the image you project through your actions, words, and style. If you do not take control of this process, then people will see and define you the way they want to, often to your detriment. You might think that being consistent with this image will make others respect and trust you, but in fact it is the opposite—over time you seem predictable and weak. Consistency is an illusion anyway—each passing day brings changes within you. You must not be afraid to express these evolutions. The powerful learn early in life that they have the freedom to mold their image, fitting the needs and moods of the moment. In this way, they keep others off balance and maintain an air of mystery. You must follow this path and find great pleasure in reinventing yourself, as if you were the author writing your own drama.

SUBVERT YOUR PATTERNS

Animals depend on instincts and habits to survive. We as humans depend on our conscious, rational thinking, which gives us greater freedom of action, the ability to alter our behavior according to circumstance. And yet that animal part of our own nature, that compulsion to repeat the same things, tends to dominate our way of thinking. We succumb to mental patterns, which makes our actions repetitive as well. This was the problem that the great architect Frank Lloyd Wright was obsessed with, and he came up with a powerful solution.

As a young architect in the 1890s, Wright could not understand why most people in his profession chose to design buildings based on patterns. Houses had to follow a certain model, determined by materials and cost. One style became popular, and people copied it endlessly. Living in such a house or working in such offices would make people feel soulless, like cogs in a machine. In nature, no two trees are ever the same. A forest is formed in a kind of random fashion and that is its beauty. Wright was determined to follow this organic model rather than the mass-produced model of the machine age. Despite the cost and energy, he decided that no two buildings of his would ever be the same in any way. He would extend this to his own behavior and interactions with others—he took delight in being capricious, in doing the opposite of what colleagues and clients expected from him. This eccentric manner of working led to the creation of revolutionary designs that made him the most famous architect of his time.

In 1934 he was commissioned by Edgar Kaufmann, a Pittsburgh department-store magnate, to design a vacation house facing a waterfall on Bear Creek in rural Pennsylvania. Wright needed to see the design in his mind before he could commit it to paper; for this project nothing would come to him, and so he decided to play a game on himself. He simply ignored the work. Months went by. Finally Kaufmann had had enough and telephoned Wright—he demanded to see the design. Wright exclaimed that it was finished. Kaufmann said he would be over in two hours to look it over.

Wright's associates were aghast—he had not yet penciled in one line. Nonplussed and with a rush of creative energy, he began to design the house. It would not face the waterfall, he decided, but stand over and incorporate it. When Kaufmann saw the design, he was delighted. The house became known as Fallingwater, often considered Wright's most beautiful creation. In essence, Wright had forced his mind to face the problem without research or preconceptions, completely in the moment. It was an exercise to free himself from prior habits and create something totally new.

What often prevent us from using the mental fluidity and freedom that we naturally possess are the physical routines in our lives. We see the same people and do the same things, and our minds follow these patterns. The solution then is to break this up. For instance, we could deliberately indulge in some random, even irrational act, perhaps doing the very opposite of what we would normally do in our day-to-day life. By taking an action we have never done before, we place ourselves in unfamiliar territory—our minds naturally awaken to the novel situation. In a similar vein, we can force ourselves to take different routes, visit strange places, encounter different people, wake up at odd hours, or read books that challenge our minds instead of dull them. We should practice this when we feel particularly blocked and uncreative. In such moments, it is best to be ruthless with ourselves and our patterns.

CREATE A SENSE OF DESTINY

In the year 1428, soldiers stationed at the garrison at the French town of Vaucouleurs began to receive visits from a sixteen-year-old girl named Jeanne d'Arc (Joan of Arc). She was the daughter of lowly peasants from a nearby poor village, and she repeated to these soldiers the same message: she had been chosen by God to rescue France from the desperate state it had fallen into. In the previous few years, the country had become overrun by English invaders, who now held the French king hostage in England. The English were on the verge of conquering the key French city of Orléans. The Dauphin, heir to the French throne, was languishing away at a castle in the country, choosing to do nothing. Jeanne had had visions from various saints who explained to her precisely what she must do—convince the Dauphin to give her troops to lead to Orléans, defeat the English there, and then lead the Dauphin to Reims, where he would be crowned the new king of France, to be known as Charles VII.

Many people in France at the time were having such visions, and the soldiers who listened to Jeanne could not help but feel skeptical. But Jeanne was different from the others. Despite the soldiers' lack of interest, she kept returning with her usual message. Nothing could discourage her. She was fearless, moving unescorted among so many restless soldiers. She spoke plainly, like any peasant girl, but there was not a shred of doubt in her voice, and her eyes were lit up with conviction. She was certain of these visions and would not rest until she had fulfilled her destiny. Her explanations of what she would do were so detailed that they seemed to have the weight of reality.

And so a few soldiers came to believe she was for real and set in motion a chain of events. They convinced the local governor to allow them to escort her to the Dauphin. The Dauphin eventually believed her story as well and gave her the troops she requested. The citizens of Orléans, convinced she was destined to be their savior, rallied to her side and helped her defeat the English. The momentum she brought to the

French side continued for well over a year, until she was captured and sold to the English and, after a lengthy trial, burned at the stake as a witch.

The story of Jeanne d'Arc demonstrates a simple principle: the higher your self-belief, the more your power to transform reality. Having supreme confidence makes you fearless and persistent, allowing you to overcome obstacles that stop most people in their tracks. It makes others believe in you as well. And the most intense form of self-belief is to feel a sense of destiny impelling you forward. This destiny can come from otherworldly sources or it can come from yourself. Think of it in these terms: you have a set of skills and experiences that make you unique. They point towards some life task that you were meant to accomplish. You see signs of this in the predilections of your youth, certain tasks you were naturally drawn to. When you are involved in this task, everything seems to flow more naturally. Believing you are destined to accomplish something does not make you passive or unfree, but the opposite. You are liberated of the normal doubts and confusions that plague us. You have a sense of purpose that guides you but does not chain you to one way of doing things. And when your willpower is so deeply engaged, it will push you past any limits or dangers.

BET ON YOURSELF

It is always easy to rationalize your own doubts and conservative instincts, particularly when times are tough. You will convince yourself that it is foolhardy to take any risks, that it is better to wait for when circumstances are more propitious.

But this is a dangerous mentality. It signifies an overall lack of confidence in yourself that will carry over to better times. You will find it hard to rouse yourself out of your defensive posture. The truth is that the greatest inventions and advances in technology or business generally come in negative periods because there is a greater necessity for creative thinking and radical solutions that break with the past. These are moments that are ripe for opportunity. While others retrench and retreat, you must think of taking risks, trying new things, and looking at the future that will come out of the present crisis.

You must always be prepared to place a bet on yourself, on your future, by heading in a direction that others seem to fear. This means you believe that if you fail, you have the inner resources to recover. This belief acts as a kind of mental safety net. When you move ahead on some new venture or direction, your mind will snap to attention; your energy will be focused and intense. By making yourself feel the necessity to be creative, your mind will rise to the occasion.

Reversal of Perspective

For most of us, the words "ego" and "egotism" express something negative. Egotistical people have an oversize opinion of themselves. Instead of considering what is important for society, a group, or family, they think first and foremost of themselves and act upon this. Their vision is narrowed to the point of seeing everything in reference to their needs and desires.

But there is another way to look at it: we all have an ego, a sense of who we are. And this ego, or self-relationship, is either strong or weak.

People with a weak ego do not have a secure sense of their worth or potential. They pay extra attention to the opinions of others. They might perceive anything as a personal attack or affront. They need constant attention and validation from others. To compensate for and disguise this fragility, they will often assume an arrogant, aggressive front. This needy, dependent, self-obsessed variety of ego is what we find irritating and distasteful.

A strong ego, however, is completely different. People who have a solid sense of their own value and who feel secure about themselves have the capacity to look at the world with greater objectivity. They can be more considerate and thoughtful because they can get outside of themselves. People with a strong ego set up boundaries—their sense of pride will not allow them to accept manipulative or hurtful behavior. We generally like to be around such types. Their confidence and strength is contagious. To have such a strong ego should be an ideal for all of us.

So many people who attain the heights of power in this culture—celebrities, for instance—have to make a show of false humility and modesty, as if they got as far as they did by accident and not by ego or ambition. They want to act as if they are no different from anyone else and are almost embarrassed by their power and success. These are all signs of a weak ego. As an egotist of the strong variety, you trumpet your individuality and take great pride in your accomplishments. If others cannot accept that, or judge you as arrogant, that is their problem, not yours.

WE ARE FREE WHEN OUR ACTS PROCEED FROM OUR ENTIRE PERSONALITY, WHEN THEY EXPRESS IT, WHEN THEY EXHIBIT THAT INDEFINABLE RESEMBLANCE TO IT WHICH WE FIND OCCASIONALLY BETWEEN THE ARTIST AND HIS WORK.

—Henri Bergson

CHAPTER TEN

Confront
Your Mortality—
the Sublime

IN THE FACE OF OUR INEVITABLE MORTALITY WE CAN
DO ONE OF TWO THINGS. WE CAN ATTEMPT TO AVOID
THE THOUGHT AT ALL COSTS, CLINGING TO THE ILLU-
SION THAT WE HAVE ALL THE TIME IN THE WORLD. OR
WE CAN CONFRONT THIS REALITY, ACCEPT AND EVEN
EMBRACE IT, CONVERTING OUR CONSCIOUSNESS OF
DEATH INTO SOMETHING POSITIVE AND ACTIVE. IN
ADOPTING SUCH A FEARLESS PHILOSOPHY, WE GAIN
A SENSE OF PROPORTION, BECOME ABLE TO SEPA-
RATE WHAT IS PETTY FROM WHAT IS TRULY IMPORTANT.
KNOWING OUR DAYS TO BE NUMBERED, WE HAVE A
SENSE OF URGENCY AND MISSION. WE CAN APPRECI-
ATE LIFE ALL THE MORE FOR ITS IMPERMANENCE. IF
WE CAN OVERCOME THE FEAR OF DEATH, THEN THERE
IS NOTHING LEFT TO FEAR.

The Hustler's Metamorphosis

I HAD REACHED THE POINT AT WHICH I WAS *NOT AFRAID TO DIE*. THIS SPIRIT MADE ME A FREEMAN IN *FACT*, WHILE I REMAINED A SLAVE IN *FORM*.

—Frederick Douglass

By the mid-1990s Curtis Jackson felt supremely dissatisfied with his life as a hustler. The only way up and out that he could see was music. He had some talent as a rapper, but that wouldn't get him very far in this world. He felt somewhat confused about how to break into the business, and he was impatient to begin the process. Then one evening in 1996 all of that changed: at a Manhattan nightclub Curtis (now known as 50 Cent) met the famous rapper and producer Jam Master Jay. He could sense that this was his one opportunity, and he would have to make the best of it. He talked Jay into letting him visit his studio the following day to hear him rap. There he managed to impress him enough that Jay agreed to serve as his mentor. It seemed that everything now would fall into place.

Fifty had saved money to tide him over while he moved into this new career, but it wouldn't last forever. Jay got him a few gigs, but they didn't pay. On the streets near his home he would see his hustler friends doing well, while his funds were dwindling to nothing. What would he do when he ran out of money? He had already sold his car and jewelry. He had recently fathered a son with his girlfriend and he needed money to support the child. He started to feel more impatient than ever. After much persistence he got someone at Columbia Records to hear his music, and the label became interested in signing him to a deal. But to get out of a contract he had signed with Jay, he had to give him almost all of the advance money from Columbia. What was worse, at Columbia he now found himself lost amid all the other rappers signed to the label. His future looked more uncertain than ever.

With his savings almost gone, he would now have to return to hustling on the streets, and this made him feel bitter. His former colleagues were not too happy to see him again. Feeling like he needed money fast, he became more aggressive than usual and made some enemies on the streets who began to threaten him. He had been splitting his time between the record studio and hustling, and his first album at Columbia was about to come out, but the label was doing nothing to promote it. Everything in his life seemed to be unraveling at the same time.

Then one afternoon in May of 2000, as he got into the backseat of a friend's car, a young man suddenly appeared at the car window brandishing a gun and began firing at him at close range. The bullets went all over, nine of them piercing his body, including one that opened a giant hole in his jaw. The assassin then hurried away to an awaiting car, certain he had done his job with the shot to Fifty's head. Fifty's friends quickly drove him to the nearest hospital. As it all unfolded, the event itself didn't appear real to him. It was like a movie, something he had seen happen to others. But at one point, while he was being operated on, he sensed that he was close to death and suddenly it all seemed very real. A searing light flooded his eyes, and for a few seconds a shadow crept over it, while everything else came to a stop. It was an oddly calm moment; then it passed.

In the months to come he would stay in his grandparents' house, recovering from the near mortal wounds he had suffered. As he regained his strength he could almost laugh at the whole thing. He had cheated death. Of course, for hustlers in the hood this was no big deal and nobody would feel sorry for him. He had to move on and not look back, while also keeping an eye on the killers who would be looking to finish the job. In the wake of the shooting, Columbia had canceled his album and dropped him from the label—he was surrounded by too much violence. Fifty would get his revenge—he would launch the kind of mixtape campaign on the streets that would make him famous and those same executives would come back, begging to sign him.

As he geared up for action, however, he noticed that something had changed inside him. He found himself getting up earlier than usual in the morning and writing songs late into the night, completely immersed in his work. When it came to distributing his tapes on the streets, he didn't care about making money in the present—he couldn't care less anymore about clothes, jewelry, or nightlife. Every penny he made he poured back into the campaign. He didn't pay attention to all of the petty squabbles others were trying to drag him into. His eyes were focused on one thing alone, and nothing else mattered. Some days he would work with an intensity that surprised him. He was putting everything he had into this one shot at success—there was no plan B.

In the back of his mind he knew that it was that moment near death that had changed him for good. He could still feel the original sensation in his body, the light and the shadow, and it filled him with a sense of urgency he had never experienced before, as if death were on his heels. In the months before the shooting, everything had been falling apart; now it was all falling into place, like destiny.

Years later, as he amassed his business empire, Fifty began to encounter more and more people playing strange power games. A company that had partnered with him would suddenly want to renegotiate their contract or act skittish and consider pulling out—acting as if they had just found out about his notorious past. Perhaps it was just a ruse to squeeze out better terms. Then there were those at his record label who treated him with increasing disrespect and offered him meager publicity or marketing money, in a take-it-or-leave-it ploy. Finally there were those who had worked for him from the beginning, but now, smelling money from his success, began to make unreasonable demands.

Certain things mattered to him more than anything else—maintaining his long-term mobility, working with those who were excited and not mercenary, controlling his image and not muddying it for the sake of quick money. What this translated into was simple: he would exercise his power to walk away from any situation or person that compromised these values. He would tell the company trying to renegotiate terms that he was no longer interested in working with them. With the record label, he would ignore the ploy and pour his own money into the marketing of his album, with the idea of leaving them soon and striking out on his own. He would cut loose the former friends, without a second thought.

In his experience, whenever he felt as if he had too much to lose and he held on to others or to deals out of fear of the alternative, he ended up losing a lot more. He realized that the key in life is to *always* be willing to walk away. He was often surprised that in doing so, or even feeling that way, people would come back to him on his terms, now fearing what they might lose in the process. And if they didn't return, then good riddance.

If he had thought about it at the time, he would have realized that turning his back in this way was an attitude and philosophy that had crystallized in his mind that afternoon of the shooting, when death had brushed against him. Clinging to people or situations out of fear is like desperately holding on to life on even the worst terms, and he had now moved far beyond such a point. He was not afraid of death, so how could he be afraid of anything anymore?

‖‖

The Fearless Approach

‖‖‖

PEOPLE TALK ABOUT MY GETTING SHOT LIKE IT REPRESENTED SOMETHING SPECIAL. THEY ACT LIKE THEY'RE NOT FACING THE SAME THING. BUT SOME DAY EVERYBODY HAS TO FACE A BULLET WITH HIS OR HER NAME ON IT.

−50 Cent

With the language skills that our primitive ancestors developed, we humans became rational creatures, gaining the ability to look into the future and dominate the environment. But with this good came a bad that has caused us endless suffering—unlike any other animal, we are conscious of our mortality. This is the source of all our fears. This consciousness of death is nothing more than a thought of the future that awaits us, but this thought is associated with intense pain and separation. It comes with an attendant thought that occasionally haunts us—what good is it to work so hard, defer immediate pleasures, and accumulate money and power, if one day, perhaps tomorrow, we die? Death seems to cancel out all our efforts and make things meaningless.

If we were to give ourselves up to these two trains of thought—the pain and the meaninglessness—we would almost be paralyzed into inaction or driven to suicide. But consciously and unconsciously we invented two solutions to this awareness. The most primitive was the creation of the concept of an afterlife that would alleviate our fears and give our actions in the present much meaning. The second solution—the one that has come to dominate our thinking in the present—is to attempt to forget our mortality and bury ourselves in the moment. This means actively repressing any thought of death itself. To aid in this, we distract our minds with routines and banal concerns. Occasionally we are reminded of our fear when someone close to us dies, but generally we have developed the habit of drowning it out with our daily concerns.

The problem, however, is that this repression is not really effective. We generally become conscious of our mortality at the age of four or five. At that moment, such a thought had a profound impact on our psyches. We associated it with feelings of separation from loved ones, with any kind of darkness, chaos, or the unknown. And it troubled us deeply. This fear has sat inside of us ever since. It is impossible to completely eradicate or avoid such an immense thought; it sneaks in through another door, seeps into our behavior in ways we cannot even begin to imagine.

Death represents the ultimate reality—a limit to our days and efforts in a definitive fashion. We have to face it alone and leave behind all that we know and love—a complete separation. It is associated with physical and mental pain. To repress the thought, we must then avoid anything that reminds us of death. We therefore indulge in all kinds of fantasies and illusions, struggling to keep out of our minds any kind of hard and unavoidable reality. We cling to jobs, relationships, and comfortable positions, all to elude the feeling of separation. We grow overly conservative because any kind of risk might entail adversity, failure, or pain. We keep ourselves surrounded by others to drown out the thought of our essential aloneness. We may not be consciously aware of this, but in the end we expend an intense amount of psychic energy in these repressions. The fear of death does not go away; it merely returns in smaller anxieties and habits that limit our enjoyment of life.

There is a third and fearless way, however, to deal with mortality. From the moment we are born, we carry inside ourselves our death. It is not some outside event that ends our days but something within us. We have only so many days to live. This amount of time is something unique to us; it is ours alone, our only true possession. If we run away from this reality by avoiding the thought of death, we are really running away from ourselves. We are denying the one thing that cannot be denied; we are living a lie. The fearless approach requires that you accept the fact that you have only so much time to live, and that life itself inevitably involves levels of pain and separation. By embracing this, you embrace life itself and accept everything about it. Depending on a belief in an afterlife or drowning yourself in the moment to avoid pain is to despise reality, which is to despise life itself.

When you choose to affirm life by confronting your mortality, everything changes. What matters to you now is to live your days well, as fully as possible. You could choose to do this by pursuing endless pleasures, but nothing becomes boring more quickly than having to always search for new distractions. If attaining certain goals becomes your greatest source of pleasure, then your days are filled with purpose and direction, and whenever death comes, you have no regrets. You do not fall into nihilistic thinking about the futility of it all, because that is a supreme waste of the brief time you have been given. You now have a way of measuring what matters in life—compared to the shortness of your days, petty battles and anxieties have no weight. You have a sense of urgency and commitment—what you do you must do well, with all of your energy, not with a mind shooting off in a hundred directions.

To accomplish this is remarkably simple. It is a matter of looking inward and seeing death as something that you carry within. It is a part of you that cannot be repressed. It does not mean that you brood about it, but that you have continual awareness of a reality that you come to embrace. You convert the terrified, denial-type relationship to death into something active and positive—finally released from pettiness, useless anxieties, and fearful, timid responses.

This third, fearless way of approaching death originated in the ancient world, in the philosophy known as Stoicism. The core of Stoicism is learning the art of how to die, which paradoxically teaches you how to live. And perhaps the greatest Stoic writer in the ancient world was Seneca the Younger, born around 4 B.C. As a young man, Seneca was an extremely gifted orator, which led to a promising political career. But as part of a pattern that would continue throughout his life, this gift incurred the envy of those who felt inferior.

In A.D. 41, with trumped-up charges from an envious courtier, Emperor Claudius banished Seneca to the island of Corsica, where he would languish essentially alone for eight long years. Seneca had been familiar with Stoic philosophy, but now on this barely inhabited island he would have to practice it in real life. It was not easy. He found himself indulging in all kinds of fantasies and falling into despair. It was a constant struggle, reflected in his many letters to friends back in Rome. But slowly he conquered all of his fears by first conquering his fear of death.

He practiced all kinds of mental exercises, imagining painful forms of death and possible tragic endings. He would make them familiar and not frightening. He used a sense of shame—to fear his mortality would mean he abhorred nature itself, which decreed the death of all living things, and that would mean he was inferior to the smallest animal that accepted its death without complaint. Slowly he extirpated this fear and felt a sense of liberation. Feeling that he had a mission to communicate this newfound power of his to the world, he wrote at a furious pace.

In A.D. 49 he was finally exonerated, recalled to Rome, and appointed to a high position as praetor and private tutor to the twelve-year-old boy Lucius Domitius Ahenobarbus (soon to be known as Emperor Nero). During the first five years of Nero's reign, Seneca was the de facto ruler of the Roman Empire, as the young Emperor gave himself over to the pleasures that were to later dominate his life. Seneca had to constantly struggle to rein in some of Nero's violent tendencies, but for the most part those years were prosperous and the empire was well governed. Then envy set in again, and Nero's courtiers began spreading stories that Seneca was enriching himself at the expense of the state. By A.D. 62, Seneca could see the writing on the wall, and he retired from public life to a country house, handing over almost all of his wealth to Nero. In A.D. 65 he was implicated in a plot to kill the emperor, and an officer was sent to, in the Roman style, order Seneca to kill himself.

He calmly asked permission to review his will. This was refused. He turned to his friends who were present and said, "Being forbidden to show gratitude for your services, I leave you my one remaining possession and my best: the pattern of my life." Now he would be reenacting what he had rehearsed in his mind so many years before. His ensuing suicide was horrifically difficult—he sliced the veins in his arms and ankles, sat in a hot bath to make the blood flow faster, and even drank poison. The death was slow and incredibly painful, but he maintained his calmness to the end, making sure that everyone would see that his death matched his life and his philosophy.

As Seneca understood, to free yourself from fear you must work backward. You start with the thought of your mortality. You accept and embrace this reality. You think ahead to the inevitable moment of your death and determine to face it as bravely as possible. The more you contemplate your mortality, the less you fear it—it becomes a fact you no longer have to repress. By following this path, you know how to die well, and so you can now begin to teach yourself to live well. You will not cling to things unnecessarily. You will be strong and self-reliant, unafraid to be alone. You will have a certain lightness that comes with knowing what matters—you can laugh at what others take so seriously. The pleasures of the moment are heightened because you know their impermanence and you make the most of them. And when your time to die comes, as it will some day, you will not cringe and cry for more time, because you have lived well and have no regrets.

||||||||||||||||||||||||||||||||||||||||||||||||||||||||||||||||||||||||||||||||||||||||||||||||||

Keys to Fearlessness

||||||||||||||||||||||||||||||||||||||||||||||||||||||||||||||||||||||||||||||||||||||||||||||||||

THERE SEEMS TO HOVER SOMEWHERE IN THAT DARK PART OF ALL OUR LIVES . . . AN OBJECTLESS, TIME-LESS, SPACELESS ELEMENT OF PRIMAL FEAR AND DREAD, STEMMING, PERHAPS, FROM OUR BIRTH . . . A FEAR AND DREAD WHICH EXERCISES AN IMPELLING INFLUENCE UPON OUR LIVES. . . . AND, ACCOMPANY-ING THIS *FIRST FEAR*, IS, FOR THE WANT OF A BETTER NAME, A REFLEX URGE TOWARD ECSTASY, COMPLETE SUBMISSION, AND TRUST.

—Richard Wright

In the past, our relationship to death was much more physical and direct. We would routinely see animals killed before our eyes—for food or sacrifices. During times of plague or natural disasters we would witness countless deaths. Graveyards were not hidden away but would occupy the center of cities or adjoin churches. People would die in their homes, surrounded by friends and families. This nearness of death increased the fear of it but also made it seem more natural, much more a part of life. To mediate this fear, religion would play a powerful and important role.

The dread of death, however, has always remained intense, and with the waning of the power of religion to soothe our anxieties, we found it necessary to create a modern solution to the problem—we have almost completely banished the physical presence of death. We do not see the animals being slaughtered for our food. Cemeteries occupy outlying areas and are not part of our consciousness. In hospitals, the dying are cloistered from sight, everything made as antiseptic as possible. That we are not aware of this phenomenon is a sign of the deep repression that has taken place.

We see countless images of death in movies and in the media, but this has a paradoxical effect. Death is made to seem like something abstract, nothing more than an image on the screen. It becomes something visual and spectacular, not a personal event that awaits us. We may be obsessed with death in the movies we watch, but this only makes it harder to confront our mortality.

Banished from our conscious presence, death haunts our unconscious in the form of fears, but it also reaches our minds in the form of the Sublime. The word "sublime" comes from the Latin, meaning up to the threshold or doorway. It is a thought or experience that takes us to the threshold of death, giving us a physical intimation of this ultimate mystery, something so large and vast it eludes our powers of description. It is a reflection of death in life, but it comes in the form of something that inspires awe. To fear and avoid our mortality is debilitating; to experience it in the Sublime is therapeutic.

Children have this encounter with the Sublime quite often, particularly when confronted with something too vast and incomprehensible for their understanding—darkness, the night sky, the idea of infinity, the sense of time in millions of years, a strange sense of affinity with an animal, etc. We too have these moments in the form of any intense experience that is hard to express in words. It can come to us in moments of extreme exhaustion or exertion, when our bodies are pushed to the limit; in travel to some unusual place; or in absorbing a work of art that is too packed with ideas or images for us to process rationally. The French call an orgasm *"le petit mort,"* or little death, and the Sublime is a kind of mental orgasm, as the mind becomes flooded with something that is too much or too different. It is the shadow of death overlapping our conscious minds, but inspiring a sense of something vital and even ecstatic.

Understand: to keep death out, we bathe our minds in banality and routines; we create the illusion that it is not around us in any form. This gives us a momentary peace, but we lose all sense of connection to something larger, to life itself. We are not really living until we come to terms with our mortality. Becoming aware of the Sublime around us is a way to convert our fears into something meaningful and active, to counter the repressions of our culture. The Sublime in any form tends to evoke feelings of awe and power. Through awareness of what it is, we can open our minds to the experience and actively search it out. The following are the four sensations of a sublime moment and how to conjure them.

THE SENSE OF REBIRTH

Growing up in the suburbs of Chicago at the turn of the twentieth century, Ernest Hemingway felt completely suffocated by all the conformity and banality of life there. It made him feel dead inside. He yearned to explore the wider world, and so in 1917, at the age of eighteen, he volunteered as an ambulance driver for the Red Cross in Italy, at one of the war fronts. There he felt himself oddly impelled towards death and danger. In one incident he was nearly killed by exploding shrapnel, and the experience forever altered his way of thinking. "I died then. . . . I felt my soul or something coming right out of my body, like you'd pull a silk handkerchief out of a pocket by one corner." This feeling remained in the back of his mind for months and years to come, and it was oddly exhilarating. Surviving death in this way made him feel like he was reborn inside. Now he could write of his experiences and make his work vibrate with emotion.

This feeling, however, would fade. He would be forced into some boring journalistic job or the routines of married life. That inner deadness returned and his writing would suffer. He needed to feel that closeness to death in life again. To do so, he would have to expose himself to new dangers. This meant reporting on front-line activity in the Spanish Civil War, and later covering the bloodiest battles in France in World War II—in both cases going beyond reporting and involving himself in combat. He took up bullfighting, deep-sea fishing, and big-game hunting. He would suffer innumerable auto and airplane accidents, but that would only spur his need for more risk. Out of each experience, that sensation of being sparked back to life would return, and he could find his way to yet another novel.

This feeling of having your soul pulled out of your body like a handkerchief is the essence of a sublime sensation. For Hemingway it could be conjured only by something extreme, by a brush with death itself. We, however, can feel the sensation and its reviving benefits in smaller doses. Whenever life feels particularly dull or confining, we can force ourselves to leave familiar ground. This could mean traveling to some particularly exotic location, attempting something physically challenging (a sea voyage or scaling a mountain), or simply embarking on a new venture in which we are not certain we can succeed. In each case we are experiencing a moment of powerlessness in the face of something large and overwhelming. This feeling of control slipping out of our hands, however short and slight, is a brush with death. We may not make it; we have to raise our level of effort. In the process, our minds are exposed to new sensations. When we finish the voyage or task and come to safe ground, we feel as if we are reborn. We felt that slight pull of the handkerchief; we now have a heightened appreciation for life and a desire to live it more fully.

THE SENSE OF EVANESCENCE AND URGENCY

The first half of the fourteenth century in Japan was a time of intense turmoil—palace coups and civil wars turned the country upside down. Those of the educated classes felt particularly disturbed by this chaos. In the midst of all this revolution, a low-ranking palace poet later known as Kenkō decided to take his vows and become a Buddhist monk. But instead of retiring to a monastery, he remained in the capital, Kyoto, and quietly observed life around him as the country seemed to fall apart.

He wrote a series of short pieces that were not published in his lifetime but were later collected and printed under the name *Essays in Idleness*, the fame of this book increasing with time. Many of his observations centered on death, which was all too present in that period. But his thoughts around death went the opposite direction of brooding and morbidity. He found in them something pleasurable and even ecstatic. For instance, he pondered the evanescence of beautiful things such as cherry blossoms or youth itself. "If man were never to fade away like the dews of Adashino, never to vanish like the smoke over Toribeyama, but lingered on forever in the world, how things would lose their power to move us! The most precious thing in life is its uncertainty." This made him think of insects that lived for only a day or a week and yet how crowded such time could be. It is the shadow of death that makes everything poignant and meaningful to us.

Kenkō continually found new ways to measure the vastness of time, as it stretches into eternity. A man was buried one day in a cemetery in view of Kenkō's residence in Kyoto, the grave marker surrounded by grieving members of the family. As the years went by, he wrote, they would come less and less often, their feelings of sorrow slowly fading away. Within a span of time they would all be dead, and with them the memory of the man they had buried. The grave marker would become largely covered by grass. Those who would pass by centuries later would see it as a weird mix of stone and nature. Eventually it would disappear altogether, dissolving into the earth. In the face of this undeniable reality, of this eternal expanse, how can we not feel the preciousness of the present? It is a miracle to be alive even one more day.

There are two kinds of time we can experience—the banal and the sublime variety. Banal time is extremely limited in scope. It consists of the present moment and stretches out to a few weeks ahead of us, occasionally farther. Locked in banal time, we tend to distort events—we see things as being far more important than they are, unaware that in a few weeks or a year, what stirs us all up will not matter. The sublime variety is an intimation of the reality of the utter vastness of time and the constant changes that are going on. It requires that we lift our heads out of the moment and engage in the kinds of meditations that obsessed Kenkō. We imagine the future centuries from now or what was happening in this very spot millions of years ago. We become aware that everything is in a state of flux; nothing is permanent.

Contemplating sublime time has innumerable positive effects—it makes us feel a sense of urgency to get things done now, gives us a better grasp of what really matters, and instills a heightened appreciation of the passage of time, the poignancy and beauty of all things that fade away.

THE SENSE OF AWE

We are creatures that live in language. Everything we think and feel is framed by words—which never really quite express reality. They are merely symbols. Throughout history, people have had all kinds of unique experiences in which they witness something that exceeds the capacity to express it in words, and this elicits a feeling of awe. In 1915, the great explorer Ernest Shackleton found himself and his crew marooned on an ice floe near the continent of Antarctica. For months they floated in this desolate landscape, before managing to rescue themselves later the following year. During the time on the floe, Shackleton felt as if he were visiting the planet before humans had arrived on the scene—seeing something unchanged for millions of years—and despite the threat of death this scene represented, he felt oddly exhilarated.

In the 1960s the neurologist Oliver Sacks worked on patients who had been in a coma since the 1920s, victims of the sleeping-sickness epidemic of the time. Thanks to a new drug, they were awakened from this coma and he recorded their thoughts. He realized that they viewed reality in a much different way than anyone else did, which made him wonder about our own perception of the world—perhaps we see only a part of what is happening around us because our mental powers are determined by habits and conventions. There could be a reality we are missing. During such meditations he slipped into a sense of the Sublime.

In the 1570s, a Huguenot pastor named Jean de Léry was one of the first Westerners to live among the Brazilian tribes in the Bay of Rio. He observed all kinds of rituals that frightened him in their barbarity, but then one evening he heard tribesmen singing in a way that was so strange and unearthly, he was overwhelmed with a sudden sense of awe. "I stood there transported with delight," he later wrote. "Whenever I remember it, my heart trembles, and it seems their voices are still in my ears."

This sense of awe can be elicited by something vast or strange—endless landscapes (the sea or the desert), monuments from the distant past (the pyramids of Egypt), the unfamiliar customs of people in a foreign land. It can also be sparked by things in everyday life—for instance, focusing on the dizzying variety of animal and plant life around us that took millions of years to evolve into its present form. (The philosopher Immanuel Kant, who wrote about the Sublime, felt it in holding a swallow in his hands and gazing into its eye, feeling a strange connection between the two of them.) It can be created by particular exercises in thinking. Imagine, for example, that you had always been blind and were suddenly granted sight. Everything you saw around you would seem strange and new—the freakish form of trees, the garishness of the color green. Or try imagining the earth in its actual smallness, a speck in vast space. The Sublime on this level is merely a way of looking at things in their actual strangeness. This frees you from the prison of language and routine, this artificial world we live in. Experiencing this awe on any scale is like a sudden blast of reality—therapeutic and inspiring.

THE SENSE OF THE OCEANIC, THE CONNECTION TO ALL LIFE

In not confronting our mortality, we tend to entertain certain illusions about death. We believe that some deaths are more important or meaningful than others—that of a celebrity or prominent politician, for instance. We feel that some deaths are more tragic, coming too early or from some accident. The truth, however, is that death makes no such discriminations. It is the ultimate equalizer. It strikes rich and poor alike. For everyone, it seems to come too early and can be experienced as tragic. Absorbing this reality should have a positive effect upon us all. We share the same fate with everyone; we all deserve the same degree of compassion. It is what ultimately links all of us together, and when we look at the people around us we should see their mortality as well.

This can be extended further and further, into the Sublime—death is what links us to all living creatures as well. One organism must die so another can live. It is an endless process that we are a part of. This is what is known as an oceanic feeling—the sensation that we are not separated from the outside world but that we are part of life in all its forms.

Feeling this at moments inspires an ecstatic reaction, the very opposite of a morbid reflection on death.

||||||||||||||||||||||||||||||||||||||||||||||||||||||||||||||||||||||||||||||||||||||||||||||||||||||||||||||||||||||||||||||||||||

Reversal of Perspective

||||||||||||||||||||||||||||||||||||||||||||||||||||||||||||||||||||||||||||||||||||||||||||||||||||||||||||||||||||||||||||||||||||

In our normal perspective we see death as something diametrically opposed to life, a separate event that ends our days. As such, it is a thought that we must dread, avoid, and repress. But this is false, an idea that is actually born out of our fear. Life and death are inextricably intertwined, not separate; the one cannot exist without the other. From the moment we are born we carry our death within ourselves as a continual possibility. If we try to avoid or repress the thought, keep death on the outside, we are cutting ourselves off from life as well. If we are afraid of death, then we are afraid of life. We must turn this perspective around and face reality from within, finding a way to accept and embrace death as part of being alive. Only from such a position can we begin to overcome the fear of our mortality, and then all of the smaller fears that plague our lives.

WHEN I NEARLY DIED IT MADE ME THINK—THIS CAN HAPPEN AGAIN ANY SECOND. I BETTER HURRY AND DO WHAT I WANT. I STARTED TO LIVE LIKE I NEVER LIVED BEFORE. WHEN THE FEAR OF DEATH IS GONE, THEN NOTHING CAN BOTHER YOU AND NOBODY CAN STOP YOU.

—50 Cent

Acknowledgments

This book is dedicated to my NANA, a woman of strength, power, and great determination. She instilled in me knowledge. There is no knowledge that is not Power.

—50 Cent

First and foremost, my thanks go to Anna Biller for her loving support, her deft editing of *The 50th Law*, and her other innumerable contributions to the book.

The 50th Law owes its existence to Marc Gerald, Fifty's literary agent. He brought Fifty and me together in the first place and skillfully guided the project from start to finish. I must also thank my agent, Michael Carlisle, at InkWell Management, for his equally invaluable contributions; his assistant at Inkwell, Ethan Bassoff; and Robert Miller, publisher extraordinaire of HarperStudio, who played such an important role in shaping the concept of the book. Also at HarperStudio I would like to thank Debbie Stier, Sarah Burningham, Katie Salisbury, Kim Lewis, and Nikki Cutler; and for their work on the design of the book, Leah Carslon-Stanisic and Mary Schuck.

I would like to thank Ryan Holiday for his research assistance; Dov Charney for his support and inspiration; my good friend Lamont Jones for our many discussions on the subject; and Jeffrey Beneker, assistant professor in the incomparable Classics department at the University of Wisconsin at Madison, for his scholarly advice.

On Fifty's side, his management group, Violator, gave me tremendous support on the project. For this I must thank first and foremost Chris Lighty, CEO of Violator and the man behind the throne. Also giving generously of their time were Theo Sedlmayr, Fifty's attorney and business manager; Laurie Dobbins, president of Violator; Barry Williams, brand manager; Anthony Butler (better known as AB); Bubba; and Hov. Special mention as well goes to Joey P (co-founder of Brand Asset Digital) and to Nikki Martin, president of G-Unit Records, for her invaluable insights on Fifty from his earliest days in the business.

I would like to thank as well Tony Yayo, Busta Rhymes, Paul Rosenberg (president of Shady Records and Eminem's manager), the novelist Nikki Turner, Quincy Jones III, and Kevin and Tiffany Chiles over at DonDiva.

I would like to give special mention to George "June" Bishop for giving me the Southside tour and helping me understand the rich world of hustling.